Foundations of the Knowledge Economy

Foundations of the Knowledge Economy

Innovation, Learning and Clusters

Edited by

Knut Ingar Westeren

Nord-Trøndelag University College, Norway

Edward Elgar

Cheltenham, UK • Northampton, MA, USA

Published by
Edward Elgar Publishing Limited
The Lypiatts
15 Lansdown Road
Cheltenham
Glos GL50 2JA
UK

Edward Elgar Publishing, Inc.
William Pratt House
9 Dewey Court
Northampton
Massachusetts 01060
USA

A catalogue record for this book
is available from the British Library

Library of Congress Control Number: 2011939354

MIX
Paper from responsible sources
FSC® C018575
FSC www.fsc.org

ISBN 978 0 85793 771 1

Printed and bound by MPG Books Group, UK

Contents

Contributors

Hanas A. Cader, American University of Kuwait, Kuwait

Alexander Cordes, Lower Saxony Institute for Economic Research (NIW), Germany

Andreas P. Cornett, University of Southern Denmark, Denmark

Kjell-Åge Gotvassli, Nord-Trøndelag University College, Norway

Hans Siggaard Jensen, Aarhus University, Denmark

Jeannette K. Jones, American Intercontinental University, USA

Clarissa Pereira Junqueira, State University of the Western Parana, Brazil

John C. Leatherman, Kansas State University, USA

Gunther Maier, Vienna University of Economics and Business, Austria

Alex Murdock, London South Bank University, UK

Otto Raspe, PBL Netherlands Environmental Assessment Agency, The Netherlands

Meir Russ, University of Wisconsin – Green Bay, USA

Roland Scherer, University St Gall, Switzerland

Richard Shearmur, Université du Québec, Canada

Jefferson Andronio Ramundo Staduto, State University of the Western Parana, Brazil

Michaela Trippl, Vienna University of Economics and Business, Austria

Frank G. van Oort, Utrecht University, The Netherlands

Jan Ole Vanebo, Nord-Trøndelag University College, Norway

Manfred Walser, University St Gall, Switzerland

Knut Ingar Westeren, Nord-Trøndelag University College, Norway

1. Developments in the analysis of the knowledge economy: introductory comments

Knut Ingar Westeren

BACKGROUND

This book is called *Foundations of the Knowledge Economy: Innovation, Learning and Clusters*. The concept of the knowledge economy is one of the most used in recent economic policy debates, with roots back to the OECD report from 1996 and the Lisbon strategy from the European Union from 2000, but discussions about knowledge as a central aspect for productivity and competitiveness started much earlier. In this book we try to catch the most interesting and relevant aspects of the knowledge economy, which raises the question about focus. The first criterion for selecting the chapters is that they must link to the keywords innovation, learning and clusters. The second is that we must present something new, and the way to look at this is that the contributions shall relate to new studies which present new data and new theory developments. In this book, almost all data collections presented were done after 2005. We have also striven to publish a book with a truly international perspective. The contributors come from 11 different countries in Europe, North America, South America and the Middle East. This underlines the fact that the knowledge economy is a relevant concept in parts of the world other than Europe and North America.

How the concept 'the knowledge economy' can be defined is widely discussed, see among others OECD (1996), Smith (2002) and Cooke (2002). In this book you find definitions and interpretations of this concept in most chapters. Another question is how to measure empirically the knowledge economy in a way to characterize the knowledge 'content' of a region or a nation. One attempt to solve this question is done by Raspe and van Oort (2008); see also Chapter 6 in this book. Their suggestion is as follows.

We distinguish the following latent variables:

1. 'knowledge workers' with indicators: ICT sensitivity, educational level, creative class and communicative skills;

2. 'R&D' with indicators: the density of high- and medium-tech firms and the share of R&D employees;
3. 'innovativeness' with indicators: technical and non-technical innovations.

This definition is tested statistically on Dutch data and is used to answer questions about whether a knowledge-rich region grows faster than a region with a high density of clusters but low scores on knowledge indicators.

The keywords linked to the knowledge economy in the title of the book are innovation, learning and clusters, and this separates the book into the following four parts:

Part I: Theoretical considerations;
Part II: Innovation;
Part III: Learning and skills;
Part IV: Clusters and productivity.

In this book we want to present an updated discussion about innovation, learning and clusters as a part of an emerging knowledge economy together with a thorough discussion of the knowledge concept. In Parts II, III and IV we present chapters centered on a specific aspect of the knowledge economy in a format that presents an up-to-date theoretical background and in almost every chapter a case study with reasonably new data and conclusions. In this introductory chapter we give a summarized presentation of the chapters in each part of the book with the intention of presenting some key lessons showing new dimensions. We also have a challenging point of departure for this book. Where we find theoretical arguments and case study results that are in favor of new interpretations and new ways of understanding the knowledge economy we have reached a goal. But we are also looking for ways to ensure conventional arguments when that is the case.

COMMENTS TO PART I: THEORETICAL CONSIDERATIONS

When we use the term 'the knowledge economy' it is a good idea to go more in depth into the knowledge concept and that is what Hans Siggaard Jensen does in Chapter 2. Many articles and books addressing the knowledge economy begin by stating that the definition of knowledge concept is based on 'justified true belief'. In this book, we start with a chapter that suggests there is more. The title of Chapter 2 gives an interesting start of the discussion, 'the changing role of knowledge in the knowledge economy'. This indicates that the content, role and use of the knowledge concept have changed during time and from one philosophical tradition to another. For example, we find a discussion about how to understand and define tacit knowledge. In other chapters of this book we demonstrate that the analysis and conclusions we can draw from a research project depend on the definition chosen. One conclusion from Chapter 2 is that 'what is interesting in understanding the use of the concept of knowledge is the various types and forms and how they interact or do not interact.'

Chapter 3 is about growth of the knowledge-based economy analysed in a two-person non-cooperative game theory setting. This chapter contains a more fundamental analysis asking the question under which circumstances the knowledge-based economy will grow. A game theory model is used to explain the growth of the knowledge economy by looking at the employees in different occupational groups and how the players in the context of self-interest maximize their utility. The analysis shows how different Nash equilibriums can be achieved depending on the worker's position and the pay-off. An empirical illustration is presented using salary data for the US states of Massachusetts and Rhode Island in 2005 and 2008. The main result is that the game theory approach to the study of a knowledge-based economy shows that competition between workers for better economic opportunities leads to more knowledge-intense occupations. Both the theoretical and empirical analysis confirm this. It is interesting and important to notice that it is a built-in feature in the economy of the two states that the market economy combined with personal utility-maximizing strategies lead to a growth of the knowledge economy as explained.

COMMENTS TO PART II: INNOVATION

Part II of this book deals with innovation and consists of five chapters. In Table 1.1 we present some keywords from the chapters in Part II to give an understanding of the themes that are discussed.

Innovation has been and will be an important concept in theoretical discussions as well as in policy debates. In his book *Freefall*, Stiglitz (2010) argues that neoclassical economics can not explain innovations, a view that is shared by others like Cooke (2002). One alternative school of economic theory, evolutionary economics, building on Nelson and Winter (1982), is often looked at as the better alternative, capable of explaining innovations. Cooke (2002) argues that evolutionary economics can be subdivided into four traditions (or schools), namely the Schumpeter tradition, institutional economics, Marxist political economy and the Austrian School of Hayek and others. In Part II of this book we find that the contributions directly or indirectly relate to key elements of evolutionary economics, especially the Schumpeter tradition and institutional economics.

Table 1.1 Chapter focus, case study focus and main results in Part II

	Chapter focus	Case study focus	Main results
Chapter 4: Innovation: from Schumpeter to the knowledge economy	This chapter discusses the content and development of the innovation concept from Schumpeter to the knowledge economy. It also looks at the process of converting tacit to explicit knowledge as a part of an innovation. Can we find links between innovations, knowledge and competitiveness based on a case study?	A Norwegian company, Aker Verdal, built a steel jacket (a North Sea offshore installation) between May 2008 and June 2009, weight 7000 tonnes, cost US$100 million, and it took about 500 000 work-hours to build it. Data collection focused on process innovations during the production process and the role of knowledge creation, transfer and use.	Schumpeter's analysis of finding new combinations as a part of the innovation process is confirmed, stimuli coming both from inside and from outside the system (the firm). We found empirical relationships between knowledge and innovation. We also found examples of converting tacit to explicit knowledge as a part of the innovation process.
Chapter 5: Star scientists and regional knowledge transfer	This chapter addresses science-led innovations by looking at the most famous and productive scientists in different disciplines and how their contributions are linked to the environment in which they operate. The main focus of the analysis was how star scientists embed themselves in their regions of choice by creating knowledge linkages to academic organizations, firms and policy actors.	The authors define star scientists based on data about the number of citations they had in scientific journals from 1981 to 2002 and investigate the question to what extent and in which ways these scientists engaged in regional knowledge-transfer activities. A web-based survey conducted in 2008 of 720 star scientists formed the basis for the empirical analysis.	Results show that almost all star scientists collaborate with academic institutions at the regional level. These star scientists do mainly fundamental research and they adopt various mechanisms to share their knowledge and expertise with regional firms. Star scientists contribute to regional knowledge transfer but the regions are mainly the metro regions of the world.
Chapter 6: Firm productivity in innovative urban milieus	The main question is what kind of relationship there is between firm level productivity and firms embedded in innovative and knowledge rich environments in the Netherlands. Van Oort and Raspe question	Data from 2005 are analysed empirically for manufacturing industries and business services in regions defined as Dutch municipalities. Van Oort and Raspe use a special Dutch database that enables them to	One main finding indicates that firms experience higher productivity when located in a region with a higher intensity of successful innovative firms. But the study also shows that research and development density

	Chapter focus	Case study focus	Main results
	whether knowledge characteristics of a firm's location impact its growth. The study also takes up the role of knowledge spillovers as a central driver in a knowledge economy and the role of proximity.	identify research and development activities with indicators like the successful introduction of new products and services to the market and indicators of skills of employees. In the analysis they control for firm size and sectoral membership.	and knowledge workers density in the region have no significant impact on firm productivity. They find that external innovation sources enhance individual firm productivity, but the external effect is small, 2.5% compared to the 97.5% that comes from inside the firm.
Chapter 7: Not being there: why local innovation is not (always) related to local factors	This study challenges the popular notion that firm level innovation is influenced by local environments. The study focuses on the firms' internal resources to absorb knowledge transfers. This also raises the question from where the most valuable transfers come.	Shearmur performs an analysis based on the district of Quebec in Canada which again is subdivided into 50 regions. He analyses establishments in the manufacturing sector and knowledge intensive business services based on data from 2005.	The hypothesis that establishment-level innovation varies across Quebec according to local institutional, cultural and other conditions is not supported, *unless* these also vary in a continuous fashion with distance from Quebec's larger cities. Radical product innovations have the largest interaction intensity, followed by product innovations new to the firm, then radical process innovations and finally process innovations new to the firm.
Chapter 8: Innovation and creative leadership in local government	This chapter links and combines the organizational knowledge approach to the concept of creative leadership introduced by Philip Selznick together with leadership roles and the strategic change management approach. The questions this chapter discusses are as follows:	The authors have conducted a longitudinal case study in Haram municipality, collecting data continuously from 2003 to 2011. The purpose has been to analyse and conceptualize managerial activities (leadership roles), concentrating on changing the strategic	The first conclusion is that creative leadership in local government is about balancing strategic interests and expectations of citizens, companies and other stakeholders and the capability to supplement the traditional local government paradigm and New Public Management with leadership of networks.

	Chapter focus	Case study focus	Main results
	• What is the nature of creative leadership? • How important is such leadership in solving the problems of a Norwegian municipality called Haram? • What is the linkage between such creative leadership and the knowledge economy?	orientation and the basic character of the actual case-municipality.	The key lesson then is to demonstrate that the challenges of a global economy, an increased need for higher workforce skills and a blurring of the boundaries between public and private sector all serve to enhance the growth of knowledge and creativity.

We believe that the studies in Part II bring us one step closer to a better understanding of innovation, but this does not mean that we are able to give a short, concise definition of innovation. One central aspect of innovation is the process and structure as knowledge creation and diffusion change. One intention of this book is to show how the dynamics of knowledge changes influence the circumstances under which innovations take place.

Part II shows that Schumpeter still has a great influence both for theoretical discussions and as background for interpreting empirical results. One often debated question linked to Schumpeter and the understanding of innovations is whether it is the early interpretation of Schumpeter with his emphasis on the small firm innovator coming up with new combinations, or his later large firms/strong research institutions understanding that is most relevant. Several of the chapters in Part II bring up this question and, based on an evolutionary understanding, the answer is that both interpretations are still valid. Seeing new combinations is still fundamental and in Chapter 4 we find examples of how 'single person ideas' may bring about innovations in a firm. In other chapters we see how larger innovations systems can contribute. One lesson from this book is that one must be very cautious about defining the context, the institutions and the nature of the innovation situation before drawing conclusions about relations between knowledge and innovations.

The question about innovation and proximity is also raised in several of the chapters in this part of the book. The most given answer when one looks at surveys of the literature of this question is that proximity matters, and we do not debate that this may often be the case. But it is also important to note that this is not necessarily the case in any situation (see Chapter 7 in this book), depending on the needs of the firms and institutions in question.

Part II also has a chapter that discusses innovation and creative leadership in local government. The argument behind adding this chapter to this book is to show that we have some parallel discussions in the knowledge management literature (that often more specifically deals with questions like leadership) and the more economic-oriented analysis, which is the main focus of this book. It is also interesting to notice that knowledge aspects of public government raise many of the same questions we see in the business sector.

COMMENTS TO PART III: LEARNING AND SKILLS

A central part of any debate about the knowledge economy is about learning and skills. In Chapter 2 we clarified how the concepts of knowledge, learning and skills were developed, used and interpreted and also commented on how these interpretations had changed and varied between different philosophical schools. In this part we still have an evolutionary understanding and the analysis shows how cultural background plays an important role for learning and development of skills; an overview of chapter focus, case study focus and main results is given in Table 1.2.

Chapter 11 takes up the fundamental question about skills and growth. Doreen Massey (1994) gave new inspiration to the analysis of the central/peripheral growth questions in a regional context. Chapter 11 supports in broad terms the central/peripheral development split, but not without exceptions like a positive concentration effect for the two less-skilled occupation groups in the rationalization-intensive low-tech industries, which are defined by high rates of cost-reducing innovation.

Several chapters in this book (e.g. Chapters 4, 9 and 14) deal with the transformation from tacit to explicit knowledge. These chapters use Polanyi's (1958, 1966) work as the main building block for the understanding of what tacit knowledge is. It is interesting to note the fruitfulness and relevance of Polanyi's work independent of the level of the analysis; Chapter 4 is a single firm analysis, Chapter 9 is on the regional level and Chapter 14 has a cluster focus.

Table 1.2 Chapter focus, case study focus and main results in Part III

	Chapter focus	Case study focus	Main results
Chapter 9: How to improve the region's ability to learn	This chapter focuses on a region's abilities to learn and how that can be improved. The main hypothesis is that if we introduce an in-depth understanding of learning processes	The authors report from an ongoing project in the region of Walgau, and they discuss how different tools can be used to enhance and foster processes of informal learning.	Results so far have been that understanding the mechanisms of the individual and collective process of learning as sketched in the model can help regional actors to more

	Chapter focus	Case study focus	Main results
	on the micro-level of single actor's learning, we can gain new insights into regional development approaches and also develop tools and strategies to foster learning regions.	The region Walgau is an alpine valley with 21 towns and about 73 000 inhabitants. The focus of the case study is on the use of measures like thematic networks, working groups and a regional wiki.	precisely design the framework conditions shaping such processes and to deepen our understanding of the interrelationship between regional development and the (informal) learning at the single actor's level.
Chapter 10: Managing the new knowledge workers	This chapter is about how to manage knowledge workers. This issue has many challenges in the fields of both leadership and personal management in knowledge-based organizations. Relevant questions include: • What are the problems of recruitment? • What are the principal motivational factors? • What forms of leaderships are attractive? • How can skill development be promoted in an organization?	This chapter is mainly a theoretical study and refers to different sources for examples.	An often-reached conclusion is that if an organization wants to have maximum knowledge output of the knowledge worker there is a demand for flexibility, personal developments and adaptation. But the author also indicates that organizations like firms and public sector institutions have collective goals that must not be overridden by far-reaching personal flexibility and freedom.
Chapter 11: What drives skill-biased regional employment in West Germany?	This chapter is about how skills can be linked to the economic outcome of regions in Germany when we differentiate between region type and industry sectors. An analysis of how locational factors like density, industry location and industrial specialization are driving skill-biased regional employment	The empirical part of the chapter divides workers according to skills into qualification classes by combining different measures such as years of schooling and other variables like position in the firm. The analysis utilizes panels of data from 2000 to 2006 put into a special type of regression analysis. The analysis uses	The most interesting finding is a division of labor: cities promote skilled employment growth whereas less skilled occupations are better off in peripheral areas. One interesting exception is the positive concentration effect for the two less-skilled occupation groups in the rationalization-intensive low-tech

	Chapter focus	Case study focus	Main results
	growth is the main goal of this chapter.	nine region types and 12 different industry sectors at the NUTS 3 level of West Germany.	industries which are defined by high rates of cost-reducing innovation.

COMMENTS TO PART IV: CLUSTERS AND PRODUCTIVITY

Part IV takes up the cluster aspect in a knowledge economy context, but with different points of departure; an overview of chapter focus, case study focus and main results is given in Table 1.3. A central issue discussed in Chapter 12 in relation to cluster development is the role of innovations. In this context, innovation can be looked at both as an independent variable, which means that it is a growth driver, and as a dependent variable with regard to policies stimulating innovative behavior in the cluster. In the short run, innovation and knowledge creation become necessary instruments in the process of regional adoption to industrial change as a response to challenges in the competitive environment. Chapter 12 gives a discussion about this in a Danish context. In Chapter 14 a similar discussion is given, but in a quite different context about poultry clusters in Brazil. It is interesting to see how both Danish and Brazilian experiences stress the role of knowledge and learning processes but how the experiences differ when we look at the role of the state and public sector interventions. Chapter 12 stresses the framework of the triple helix model as an important and integral part of cluster development and cluster polices.

Chapter 14 shows that the firms seem to relate to the 'Porter concept' (Porter, 1990), which means that they are competitors and at the same time they learn from each other and exchange information. The most important learning processes seem to be learning by imitation and learning by interaction. Tacit knowledge seems to play a more important part as firms develop knowledge in their organizations and to a larger degree are able to draw on a common labor pool. It is important to notice that this cluster develops its competitiveness and strength being in a rural district with limited contact and information with metro areas. On the other hand, it looks like regional competence systems play an important part in cluster-innovative initiatives.

Another issue in this part of the book is about cluster development in the knowledge economy, from regional to virtual space. This relates to the discussion about the importance of traditional physical proximity. Chapter 12 stresses the necessity of an appropriate institutional framework for cooperation as a central element in this type of cluster policy. Chapter 13 is about international virtual industry clusters (IVIC) and takes up the same idea, but is not stressing

Table 1.3 Chapter focus, case study focus and main results in Part IV

	Chapter focus	Case study focus	Main results
Chapter 12: Clusters and cluster development in the knowledge economy	A central issue in this chapter is the role of innovations in relation to cluster development. More specifically the chapter discusses whether it is possible to create a business environment that facilitates the appearance of a business community characterized by features traditionally connected with industrial districts or clusters that are both traditional and knowledge-based.	This chapter refers to an explorative study of clusters in the region of Southern Denmark. The study identified six mature, 28 budding and 27 potential clusters in the region. Ten new and growing clusters were chosen for closer analysis (done from 2007 to 2008) because they seemed promising from the point of view of developing possibilities.	For a proactive cluster-developing strategy it looks like physical proximity is not necessarily required. Virtual systems based on shared knowledge and technology become feasible. And it seems important that the concept of clusters has to be extended to cover the whole range of the business environment, i.e. the 'triple helix' framework.
Chapter 13: International virtual industry clusters and SMEs: early process policy recommendations	This chapter discusses how small and medium-sized enterprises in a knowledge-deprived region can be supported by regional or national economic development policies so they can participate effectively in an international virtual industry cluster (IVIC) as well as in the knowledge economy and not be left out or left behind.	The empirical background for this chapter is a study that developed and validated a framework and set of indicators for a knowledge-deprived, economically peripheral region, Northeast Wisconsin, USA.	The authors discuss what is of special interest for an IVIC at an early stage and they emphasize leadership and knowledge spillovers. They also recommend a process where the social capital and culture of the regional environment can be utilized positively to build local identity.
Chapter 14: The knowledge, learning and innovation process: the poultry cluster in the western region of Parana, Brazil	This chapter is about knowledge, learning and innovation processes in a poultry cluster in the western region of the state of Parana in Brazil.	The case discussed in this chapter uses data from five companies and focus is on their innovations and innovative behavior. All companies had contractual relationships with	We find only incremental innovations such as product and production innovations in the companies. It seems as though these innovations are responses to the

	Chapter focus	Case study focus	Main results
		their suppliers through vertical integration. The field research took place between December 2007 and April 2008.	need of the firms to be flexible in their market operation. The most important learning processes seem to be learning by imitation and learning by interaction. Tacit knowledge seems to play a more important role as firms develop knowledge and to a larger degree are able to draw on a common labor pool.

the role of the public sector in the same way. This can be understood by the much stronger attitude of incorporating public sector activities with firm initiatives in Scandinavian countries compared to the USA.

Chapter 13 also discusses cluster policies towards small and medium-sized enterprises in a knowledge-deprived region so that they can participate effectively in an international virtual industry cluster. The following list is used:

- by infrastructure – focusing especially on telecommunications and transport;
- by capital (venture capital);
- networking – the importance of knowledge sources are underlined;
- brand creation.

This list has many similarities with what is discussed in Chapter 12, but we must be aware of the fact that the analyses in Chapter 13 are about a knowledge-deprived region in a country (USA) that relies heavily on private markets and to a lesser degree on public policy interventions. Their analysis specifically takes into consideration that a rather poor knowledge-deprived region will never take full advantages of proximity to the rich mega-regions.

REFERENCES

Cooke, P. (2002), *Knowledge Economies: Clusters, Learning and Cooperative Advantage*, London: Routledge.

Massey, D. (1994), *Space, Place and Gender*, Minneapolis: University of Minnesota Press.

Nelson, R.R. and S.G. Winter (1982), *An Evolutionary Theory of Economic Change*, Cambridge, MA and London: Harvard University Press.

OECD (Organisation for Economic Co-operation and Development) (1996), *The Knowledge-based Economy*, Paris: OECD.

Polanyi, M. (1958), *Personal Knowledge*, Chicago: University of Chicago Press.

Polanyi, M. (1966), *The Tacit Dimension*, New York: Doubleday.

Porter, M. (1990), *The Competitive Advantage of Nations*, New York: The Free Press.

Raspe, O. and F.G. van Oort (2008), 'Firm growth and localized knowledge externalities', *Journal of Regional Analysis and Policy*, **38**, 100–16.

Smith, K. (2002), 'What is the "Knowledge Economy"? Knowledge Intensity and Distributed Knowledge Bases', The United Nations University, Discussion Paper Series 2002–6, Maastricht.

Stiglitz, J.E. (2010), *Freefall: America, Free Markets, and the Sinking of the World Economy*, New York: W.W. Norton.

PART I

Theoretical considerations

2. The changing role of knowledge in the knowledge economy: concepts of knowledge and knowledge management

Hans Siggaard Jensen

There is a long epistemological tradition attempting to understand our concept of knowledge. This tradition goes back at least to the work of Plato in the dialogue 'Theaetetus', where the first presentation is made of the analysis of knowledge as justified true belief. Most subsequent major philosophers have worked with this concept and have given different answers to the question regarding what knowledge is. But the use of the concept of knowledge has changed considerably. The emergence of a successful form of science during the seventeenth century made the results of science the dominant paradigm of knowledge. And it became clear that there is a question concerning the relation between the type of knowledge we have when we know that we have two hands, and when we know a scientific law. Of what nature this difference is exactly, is still being debated. But a large social institution – the research system – has been developed to organize the creation of scientific knowledge. The enormous and rapid growth of the industrial economies is usually ascribed to the new role of knowledge and innovation and the emergence of social institutions that could handle the presuppositions for creation of knowledge and implementations of innovation. Freedoms like free speech, a free press, the free pursuit of learning and of knowledge are here seen as important (Mokyr, 2002). The emergence of concepts such as knowledge economy, knowledge society and knowledge management signals a change of focus from knowledge per se to knowledge in use (Warsh, 2006; OECD, 1996). Furthermore, the emergence of IT systems that claim the automation of knowledge processes, such as in expert systems and knowledge-based systems, also indicates a fundamental shift in the control of the use of the concept of knowledge. Many of these developments were analysed and predicted by the French philosopher Francois Lyotard in his book *The Post-modern Condition, A Report on Knowledge* from (Lyotard, 1979). The ways in which the concept and understanding of knowledge have been approached have often been through a conceptual analysis, on the assumption that either the logical properties or the actual daily use of the concept would be the material out of which the understanding would flow. What is important though is to recognize the wide variety of uses of the concept and the widely divergent types of sources and traditions. We can actually ask whether there is such a thing as a concept of knowledge. There are probably several, and it is important to be open to the possibility that these concepts have what has been called 'family resemblance', rather than a shared common nucleus, which would then be the 'real' concept of knowledge. The use of the term 'knowledge economy' signals a change in the use of the concept from a more epistemological perspective to an economical – without sacrificing the epistemological viewpoint.

There is always a point in distinguishing between the word, the term (in the sense of the 'meaning' of the word understood as the set of words it would translate to in various languages) and the concept; and in being clear on the 'sorting' of concepts. That is when there are types or forms; there are differences in a concept (of those things falling under the concept). Sometimes there is an important significance in clarifying that although one word is used, two concepts are involved. And this may change from language to language. So what is called 'knowledge' in one language may be designated with several words in other languages, thereby revealing important differences either in the concept or between concepts. An example could maybe clarify. The word 'tree' in English is different from 'wood', but in some languages there is no such difference. This does not imply that they do not have the conceptual difference between referring to an organism and to a type of material. The reason for using the same word is of course that wood comes from trees which depend on the conceptual schemes involved. We will return to this issue later when discussing types and forms of knowledge.

The following analysis is based on the idea of looking at various conceptual domains, where the concept of knowledge is being used with a focus on these domains as established domains of expertise. It is worth noting the tendency towards a greater variety of uses and sources of significant uses of the concept. An example could be the emergence of the idea in connection with the development of 'artificial intelligence' of knowledge-based systems. As stated earlier, the term 'knowledge economy' is somewhat meaningless because in a way all economies are completely dependent on 'knowledge'. Therefore, it is important to track the various uses and the domains of usage and try to understand the various dimensions of the development of the concept of knowledge.

We may briefly create an overview of the various aspects of the concept by looking at what is seen as the 'opposite' of knowledge. Typically we have something which is not knowledge, but also not just the 'absolute' opposite, ignorance (Unger, 1975; Rescher, 2009), and then something is added to make it knowledge. The presence of something is thus it seems presupposed, and that something can also vary. The absolute opposite can be ignorance, both as the total lack of ideas or concepts or as the total lack of competence. In the case of the most traditional analysis of knowledge the analysis is based on justified true belief. The starting point is 'belief' and then something is added. With the idea of belief as a starting point, a number of things are already accepted. That knowledge is at least a certain state of mind is one idea (Steup, 2008; Greenough and Pritchard, 2009). But the starting point could be other than belief. If the focus is upon how knowledge is acquired in a process of learning then it could be experience, so that knowledge is seen as experience plus something, typically a certain process of inquiry or discussion. The pragmatist theory of knowledge could here be an example as it has developed from Charles S. Peirce to John Dewey (Peirce, 1877; Dewey, 1938). Or it could be a theory formulating what knowledge is by basing this on an understanding of what a scientific theory is, thus taking scientific knowledge as the basic form of knowledge. Such a theory could then be discussed and criticized by showing that it does not fulfil the formal or logical requirements needed or by showing that it simply is not fitting the sort of phenomena that we actually consider as science. The views of Karl Popper could here be used as examples as he requires that although scientific theories are conjectures they should fulfil certain logical requirements such as being of a form so that they are falsifiable (Popper, 1972). In the IT area the concept of knowledge is often arrived at by taking as a starting point the concept of information and then claiming that something has to be added to information to make it knowledge (Dretske, 1991). This is the substance of the so-called knowledge pyramid consisting of data, information and knowledge. The basic input here is data, which is usually understood as formalized representations of facts, such representations as a computer can manipulate. This is a purely syntactic concept, because something is data when it conforms to a syntax for a notation. Information is then a more semantic concept, data with meaning. For example, '21°C' is data in the sense that it conforms to a notation with two fields: a decimal number and an indication of the type of temperature scale. So instead of °C there could have been a °F, but not a B in the usual definition of how we represent temperature. It is information in the sense that it has meaning, and this is implied already if we say that it is data about temperature. If it is data about centuries, it has a totally different meaning; the same data gives different information. If a person knows that the temperature in the room is 21°C by looking at a digital thermometer in the room, then we would say that we are in the domain of knowledge, and ask whether we can trust the thermometer, whether it functions, and so on. Looking at a thermometer in a room where the temperature is actually 21°C, but finding out that the thermometer stopped working yesterday, will tell us that we did not in fact have knowledge after looking at the thermometer.

Finally, we can see knowledge in contrast to opinion. Often what is needed to transform mere opinion into knowledge is evidence. Here a social dimension is included in the sense that we form opinions based on discussion and testimony using processes of trust and persuasion. If we look at the knowledge system it is exactly through discussions and persuasions that scientific 'facts' are established. The most recent trend is towards the systematic analysis of publications in a certain field concerning a certain issue or question. Knowledge, here taken in the sense of evidence, becomes the consensus or near consensus in the literature on the answer to the question at issue. This is actually the creation of a form of meta-knowledge, where we try to get to know about the knowledge claims put forward in the scientific literature. Obviously, issues then arise as to what literature to take into account and how to assess this literature. The relation here between knowledge and evidence is very tight, to the point that they are considered the same. Scientific knowledge in this sense is then the result of a systematic review of serious knowledge claims made in the course of the working, mainly through publication, of the scientific system understood as an institution. The difference in relation to how it 'normally' works is the idea of a form of systematic judgement, rather than the more spontaneous creation of a scientific consensus, a fact (Petticrew and Roberts, 2006).

An interesting position in the theory of knowledge is the one presented by Timothy Williamson in the book 'Knowledge and its Limits' from 2000. In this book Williamson claims that the concept is so fundamental that one cannot analyse it. This is based on the argument that any analysis presented will need to use the concept. Thus, in giving accounts of what it means to believe something, one has to be able to give meaning to a claim that one knows what one believes. Or if one is going to give an analysis based on concepts like justification, perception, experience etc., one will also presuppose that one has a concept of knowledge to be able to give meaning to these concepts. This is a very forceful position based on a particular conception of what it is to give a logical analysis of the concept of knowledge. But one can also try to analyse the concept in the way it is actually being used. The same type of issue has been raised about such a fundamental concept as causation. It has proved extremely difficult to give an analysis of the concept. It is easier to try to give an account of how the concept is used in actual practice when talking about causes.[1]

The various conceptions of knowledge sketched above give rise to two types of deficiency in the lack of knowledge. If a person is ignorant, then there is a lack of knowledge, but if one is mistaken, there is a different form of deficiency. We usually call it error. Beliefs can be false, and are thus not knowledge. But we can also lack knowledge in the sense of being unable to do something. If we make knowledge claims in this case, these are false, and we lack a certain competence in the sense of an ability to do something effectively. This relates to the difference often expressed by the phrases 'knowing that' and 'knowing how' (Ryle, 1949). The first is propositional and the second dispositional. If one does not know how to ride a bike, the lack of such knowledge is not something false or erroneous, but an inability that is not necessarily total ignorance in the sense that one may not even know what a bike is or what it means to ride a bike; but the inability to effectively carry out certain practices, the successful mastering of which constitutes what it is to ride a bike. When we then make a distinction between knowledge per se and knowledge in use we focus on a distinction between having knowledge and being able to use that knowledge. This relates to a distinction between knowledge and competence. Having 'knowledge how' is clearly the ability to use that knowledge, whereas this is not the case with 'knowing that'. The interplay between 'knowing that' in the sense of knowledge per se, 'knowing how' and knowledge in use has been at the center of the conceptual development of knowledge when we talk about the knowledge economy, the knowledge society and knowledge management.[2] The idea of a knowledge-based economy goes back at least to the work of Fritz Machlup in the 1960s and was promoted heavily by the OECD in the 1990s. The emergence of the concept and the development of indicators for a knowledge-based economy have been detailed by Benoit Godin in a number of papers as part of a historical project on measurement of science and of knowledge.[3] He especially focuses on the broad empirical approach of Machlup and the influence of the OECD in shaping the idea of a broader function of knowledge in society and a wider variety of ways it could function in the economic system. The traditional way of seeing the function of knowledge was through the practical application of research results in the creation of new products or processes via innovation and development. This was of course characteristic of the science-based industrial economy that emerged at the end of the nineteenth century, when physics, chemistry and biology became foundations for new industries like electronics, pharmaceuticals and brewing. The emergence of a closer relation between university-based research, research in government knowledge institutions and research and development in private companies was typical of the advanced industrial economics and can be exemplified by the system built in the new German Imperial State under the guidance of Friedrich Althoff, often named 'the Althoff system' (Backhaus, 1993). In this system the role of basic research and a system for its application is central. The success of the German development of university research and its application in industry was of great influence and central to the understanding of an advanced industrialized economy. It also led to a focus on understanding the central phenomenon of science and of the organization of research. The conception of knowledge became one centered on the result of the research process when done in a professional and methodical way. Science and research became professions and a system was developed to secure this. One of the forms was the development of a system of research education and a system of organization for the sciences. The organization was based on the central concept of a scientific discipline, which had its own criteria of identity and quality. A discipline should have an object of study and a method for studying it. From the end of the nineteenth century, several proposals have been made to organize and characterize the various scientific disciplines and justify their claims to be producers and repositories of knowledge.

KNOWLEDGE AS A VALUE CREATOR

The focus on knowledge that we have seen in the last decades has brought new perspectives. The philosophical, scientific and IT perspectives have been augmented with perspectives from economics and management. These perspectives have again had feedback effects on more conceptual issues. Foremost in this respect has been the renewed focus on the distinction between tacit and explicit knowledge, which has come to be seen in a new light. The focus has shifted from looking at what it is to have knowledge and how it is 'stored' (in minds and in libraries) to what it is to use knowledge, especially in the creation of value. This has to do with the general development of the economy and the business system. I will look at two developments that seem to be central. The first one is the development of an economic understanding of knowledge as seen in the so-called new growth theory, and the second is the development of the field of knowledge management. Both these developments connect to fundamental issues concerning our concepts of knowledge and both have close connection with the issues concerning the development and organization of enterprises. They are also central to understanding the role of knowledge in securing productivity, competitiveness and innovation. In addition, these developments may be seen as two aspects of the knowledge economy, in the sense of a type of society where learning, research and management enter into new relations.

It has been generally recognized that knowledge in many forms, particularly in the form it has taken in highly developed industrial societies, plays a central role for economic development and an increasingly larger role in business. Science and technology have been seen as drivers in the historical development of industrialized societies as an important factor which has been responsible for their rapid and astonishing growth (Appleby, 2010). But there has been a lack of understanding of how and why this was possible. In economic theory, knowledge and technology have been seen as externalities. In management theory, knowledge and technology have been seen as something which is not to be managed or organized, but would set the conditions for organizing and managing work and capital. In the first explicit management theory of Frederick Taylor, management was understood mainly as planning, which of course is the function of knowledge in organizing production and work processes (Taylor, 1911). The neo-classical understanding of knowledge in economic growth was based on a marginalist understanding and on understanding the growth of an economy as based on the interaction of factors of production such as labor and capital. Knowledge and technology were not understood as factors of production, but rather as external factors which shaped the way the factors of production were functioning. The classical growth theory of Solow, formulated in 1957, is the best example (Solow, 1957). That knowledge plays a central role in economic growth can be inferred from the discrepancy between the actually observed growth rates and those calculated from observations of investment in labor and capital used as input in economic models. The difference between predictions based on models and what is observed is seen to be caused by exogenous factors such as science and technology, basically knowledge. Knowledge is not understood in economic terms, but as something which is introduced into the economic system, influences it and contributes to the creation of growth, but is not itself created by economic forces.

The new growth theory developed on the basis of the work of Paul Romer takes a different view (Romer, 1990). It includes knowledge as a production factor, but one that when included transforms the nature of the production system understood in economic terms. The concept of knowledge as economic is based on its differentiation from both labor and capital. It is in Romer's term a non-rival factor and non-excludable without being a 'public good' in the sense that neo-classical economists treated such goods and services (Coase, 1974). Physical goods and services are rival and excludable. If I buy a car, only I can own and drive it at a given time. If I invest capital or labor, only I can invest this particular capital or labor. But knowledge is different. It can be invested or used at the same time in many places, and it can be used again and again without being consumed. This is something that was already noticed by Karl Marx in his analysis of the labor theory of value. If the amount of value that can be extracted from a good is always limited by the value put into it by labor, then knowledge, like the binomial theorem, seems to contradict this. It took a limited amount of labor to create the theorem, but potentially it can be used again and again to create value without being consumed itself.[4]

The non-rival and non-excludable character of knowledge pertains to all forms of knowledge, all ideas or concepts that can be used in the creation of value, whether it is scientific knowledge or knowledge in the form of mastering certain procedures or practices, that is, 'know-how'. Knowledge is not necessarily non-excludable. This is only the case if it is a public good, and thus not in any way under market forces. Some institutions imply the excludability of knowledge, such as patents, copyrights, secrecy, confidentiality, and so on. The most important character of knowledge then is its non-rival character. It is this property that is essential for the role knowledge plays in creating growth; it is the central economic character of knowledge. Because of this property, knowledge in use is able to create an economy of increasing rather than decreasing returns. This is the central hypothesis of new growth theory, which was also pre-sensed by Alfred Marshall in his work on industrial economy.[5] And an economy shaped by knowledge and the use of knowledge is thus very different from an economy shaped by diminishing returns. According to Romer, such an economy will tend towards monopolistic competition and be unstable in the sense that it will not have one well-determined equilibrium state. It will also show such important tendencies as path dependencies and an increased focus on localization and tacit knowledge (Cortright, 2001).

The important point in the new growth theory is that it attempts to treat knowledge as an endogenous factor. This means that knowledge is understood in economic terms rather than something external which has a great economic impact. One can easily get the impression that there is a close connection between this shift and the clear emergence of important economic forms of technology which share the properties of knowledge, such as digital information technology and the obvious importance of work with symbols, rather than material things recognized in the 1970s when the concept of the information society/economy emerged (Porat and Rubin, 1977). The various theories of economic development could best be described as theories about scientific, technological and social development with great economical impact. The explanations of that development were not in themselves economical.[6] One can of course ask what the epistemological significance of the non-rival and non-excludable character of knowledge is. Actually, these two properties can also be found in information. If information is stored as data, then the actual bits or marks are rival and excludable (although if data can be seen or read by more than one reader or perceiver then we have non-excludability in this case). If we make a distinction between data tokens and data types, then data types are non-rival. So from this point of view it is not an inherent characteristic that distinguishes between data, information and knowledge, but rather the use context. It is the use of data or information (whatever the precise difference is) that in economic terms constitutes these as knowledge. This is independent of what types of knowledge we refer to, whether we talk about knowledge in terms of know-that, know-why or know-how or any other types of knowledge, as long as they allow for non-rivalry and non-excludability. This obviously poses specific problems for the type of knowledge, or rather knowing, that we call tacit.

The interesting phenomenon is that these economic aspects of knowledge are independent of the important epistemic characteristics of knowledge. They hold for knowledge claims, for beliefs, even for ignorance or superstitions. In the case of ignorance, it has to be formulated, represented as information[7] and be communicable. The economically interesting characteristic is that individuals act upon what they believe to be knowledge. In this sense an economic theory of knowledge resembles a social constructivist theory of knowledge. In such a theory, knowledge is understood exactly as knowledge claims which are held or adopted through processes like persuasion and negotiation. These processes claim not to presuppose that what is known has sense, reference and truth. This means that 'social' is understood in the sense that the determining factors for what is held as being known are social processes or phenomena like power and persuasion.[8] We might say that in relation to economics it might not matter whether activities are based on superstition or on knowledge in the epistemic sense. A company that produces and sells horoscopes by manipulating symbols and using highly sophisticated software can be successful, as can an engineering consulting company that, superficially seen, performs the same kind of operations and work. As organizations, the two might also be very similarly organized. Of course a basic question is whether an activity can in the long run be based on ignorance and superstition. This is an interesting evolutionary and anthropological question (Barth, 2002). Myths, rituals and occult practices might not explicitly contain what we would in epistemic terms call knowledge, but may in other ways incorporate such knowledge in tacit forms or as know-how in a narrative form, as a result of accumulated experience.

Another basic question is: what concept of knowledge is used when someone claims that something is a construction of knowledge, a representation or revelation of knowledge? One might hold that behind knowledge claims there are only processes of power and persuasion, and that such claims do not contain 'real' knowledge or truth, but are merely social constructions. But when formulating a knowledge claim, rather than just stating the semantic content as something that can be examined or entertained, using a kind of 'epoché' or suspension of epistemic attitude, one is using a concept of knowledge, even if one denies that there is any truth or epistemic justification involved. Even stating or entertaining a proposition seems to presuppose a concept of meaning, and thus at least one form of knowledge, semantic knowledge, that concerns knowing the meaning of sentences. It is also significant that when making claims of knowledge rather than some mental act, a person clearly wants to claim some special status that must imply truth. This is even the case when making claims of belief, where of course one might claim belief and at the same time doubt. It is more difficult to claim to know that p, and then at the same time express doubt about the truth of p. Then one is clearly really only believing p. From the economical point of view these epistemic distinctions do not matter. The important point here is on the basis of how people actually act. And knowledge claims, rather than expressions of belief, serve a different persuasive purpose.

KNOWLEDGE AND VALUE CREATION IN THE FIRM

In understanding firms, the competences and capabilities of such organizations have attracted increasing attention. This development has often been thought to take its starting point with the work of Edith Penrose (1959) and has developed into the resource-based theory of the firm. This has to a large extent given rise to theories of strategy. But other theoretical developments have focused on the resources of the firm emphasizing the knowledge resources. These have centered on learning and intellectual capital. Learning has here been understood as both learning in the individual and the social or organizational sense. Firms compete in terms of their ability to improve, based on the implicit or explicit use of their experience and their ability to acquire new capabilities as required by the development in their business areas. Increasingly this also holds for organizations which are not firms in the narrow sense, or maybe not even organizations, like regions and countries. They act under market conditions, where competition is a determining factor. And the factors that determine the outcome are increasingly connected to the ability to acquire and use knowledge (Jensen, 2008). Many firms also become more and more dependent upon their intellectual resources in the shape of human and intellectual capital. The emergence of the idea of intellectual capital actually sprang from the observation that companies often had significantly higher value on the market than could be seen from their valuation according to their accounts (Stewart, 1997). These various developments led to the emergence of the concept of knowledge management (Prusak, 2001). It also led to a focus on the organizational processes surrounding knowledge and learning. In this context, two issues became especially important. The first is the rediscovery of tacit knowledge and the second is the discovery of the interaction between different forms and types of knowledge in value creation.

Tacit knowledge has been an important theme in discussions about formalization and its limits. Can one make all the assumptions of a formal system explicit within the system itself? What is available for formalization and what is available for critical discussion? The answers to these questions depend on basic issues surrounding the understanding of the concept of knowledge. Then issues surface from a new angle. One such angle involved developing expert or knowledge-based systems. Here it was discovered that actually formalizing knowledge in knowledge bases was extremely difficult, especially with more basic and elementary knowledge. Naïve physics – the basic physics knowledge we all seem to have that enables us to make predictions about objects and their behavior – proved nearly impossible to formulate. This had also been clear in many developments of IT systems, when it became necessary to describe the various routines and procedures and the knowledge involved in even rather simple tasks, when such tasks were to be automated. People seemed to know what they were doing, but without being able to account for that knowledge, at least not explicitly.

In studying professionals and experts, researchers had often claimed that forms of expertise and professionalism were silent or tacit in the sense that people had capabilities they could not explicitly describe or formulate.[9] This was, as in the work of Bertil Rolf (1991), related to the work of Michael Polanyi and his ideas of tacit knowing (Polanyi, 1958).

The other source of renewed interest was the work of the Japanese management researcher Ikujiro Nonaka, who, inspired by Kyoto school (Davis, 2010), focused on the role of tacit knowing in learning and knowledge processes in organizations. The essential element in his studies of knowledge and learning processes in organizations was the various types of transformations between explicit and tacit knowledge. A famous case study was published in *Harvard Business Review* in 1992 and was in many ways a starting point for the non-technological form of knowledge management (Nonaka, 1991). On the basis of Nonaka's work a long discussion followed concerning understanding of tacit knowledge and/or knowing (Tsoukas, 2003). These contributions focused on the role of various forms or types of knowledge and knowing and their interplay and potential transformations or translations. At the same time developments in the attempts to understand value creation and accounting in companies with a strong knowledge component focused on the forms of capital involved. New forms of capital were seen as important in understanding companies in areas such as software and biotechnology. Human capital had been introduced in the 1960s and, as mentioned above, the concept of intellectual capital had emerged in the 1990s. The other 'new' forms of capital discussed were cultural and social capital. These various forms of capital and their role in value creation were seen in relation to the various forms of knowledge and competences that played an essential role in value creation. Capital, in the sense of stored potential, was located in the various different explicit and implicit routines and practices in the organization and the various stores of information that could support these, such as databases and patents. Value creation was understood as the interaction of forms of competences and capabilities in relation to forms of knowledge and knowing. Knowledge in action formed the basis for the emergence of a new understanding of leadership and management which was related to the resource-based view of firms and organizations. An essential function of management was to manage the complex interplay between these knowledge forms and resources as they were represented in individuals and the various parts of the organization. This was an important type of emergence of management challenges and new forms of management (Wallace, 2007).

When a field becomes sufficiently complex so that it is not possible to rely on implicit regulation (effectively self-regulation) then it becomes necessary to make such regulation explicit. The history of business and management has many such cases; the emergence of the large modern highly mechanized and later automated production company is one (Hounshell, 1984). In the context of a development trend where nations and regions in a globalization process come to behave more and more as entities competing on a global market, the same issues also arise for such 'organizations'. 'Organizations' is in quotes because the traditional distinctions between an organization and a market, as well as between an organization and a society, are being blurred. Regions, states, nations see themselves as if they resemble firms. This occurs at the same time as ideas of knowledge economies emerge and thus the idea of competition is based on the capability to use regional and national knowledge resources. Thus 'knowledge management' becomes an issue also at these levels (Campbell and Pedersen, 2011). Examples of an emerging discourse of competition, based on the capabilities of knowledge and knowledge use, can be found in Bill Clinton's presentation of his first economic program and the various policy statements of the European Union, related to making the EU the most competitive knowledge area in the world.[10] This also implies a blurring of the distinction between knowledge management and knowledge politics, where political leaders act more like business leaders than politicians trying to compensate for the ills created by business, the market and competition. Knowledge is still a public good, but the ability to use knowledge is of course not.

THE IMPORTANCE OF KNOWLEDGE TYPES AND FORMS

The issue of types and forms of knowledge and knowing becomes essential. The traditional bases for typologies and classifications have been based on two dimensions: the first is the various ways in which we obtain knowledge; the second is the various domains about which we have knowledge. This compares to the traditional basis for understanding a discipline, namely the method used and the object or domain studied. The distinction between logical, formal, a priori knowledge obtained through thinking or inference alone, and empirical knowledge, obtained via observation and perception, is a good example of ways in which we obtain knowledge. The distinction between causal knowledge about natural processes and interpretative knowledge about artworks and actions is an example of differences based on what knowledge is about. An essential idea in this way of thinking is that there are essential relations between what you know and the way you know about it. Abstract mathematical entities cannot be directly observed, thus mathematics is about something that necessitates a different method than the one found in the empirical sciences. Historical facts cannot be observed, so the past must be studied with a different method than the present. Fictive persons must be studied in other ways than real living persons, although there might be similarities. One position holds that the forms of knowledge are the disciplinary forms. This overlooks the various forms of tacit knowledge and knowing. Another approach is based on knowledge as related to the ability to answer types of questions. This was briefly hinted at by Ludwig Wittgenstein in his §78 of *Philosophical Investigations* (Wittgenstein, 1953). Here three different types of questions are posed, related to knowing and saying. Do you know the height of Mont Blanc? Do you know how to use the word 'game'? Do you know the sound of a clarinet? The last two questions could be answered with a 'yes' even if one was not able to actually provide an answer in the sense of answering the question in the form 'describe the sound of a clarinet' or 'state the rules for the use of the word "game"'. Any relevant answer would typically be an exemplification. If one generalizes on the basis of these three types of questions then one gets division of forms of knowledge that would be: factual or theoretical knowledge, which is explicit and can be stated; practical knowledge, which is tacit and can be exemplified and learned; and immediate knowledge, which is experienced and tacit, and can be exemplified (Molander, 1996). There are obviously intricate interactions between these forms and the enterprise of knowing and thus obtaining and using knowledge probably always involves all of them. You need to be able to read 'practical' to get theoretical knowledge, and to be able to read you need to know the letters 'immediate'. Another area where forms of knowledge have been viewed in a broad and action- and competence-oriented perspective is theories of the curriculum or didactics. A well-known typology here is a division in factual, conceptual, procedural and meta-cognitive forms of knowledge (Kratwohl, 2002). The development of such typologies reflects the experience of working with planning and assessing, in grading and also actually measuring, knowledge and knowledge processes in education. The interesting feature of these four types of knowledge is that there is a distinction between the cognitive and the meta-cognitive. This type of distinction is also found in Harry Collins' work on expertise, where meta-expertise such as that involved in evaluation and judgement on domains or activities of knowledge such as research quality is seen as very different from the first-order expertise (Collins and Evans, 2007). Meta-cognitive knowledge involves knowing about knowledge strategies and the ability to choose the right kind of approach to using or getting knowledge in various situations, and it also involves knowing about one's own knowledge capabilities. In recent educational theory this has been seen as a central competence in learning.

If we assume that typologies and classifications are not based on some natural kinds, whereas differences of forms are, then we can say that knowledge and knowing will have types based on what aspects of knowing are seen as important in a given context. In this respect, types of knowledge might then be like disciplines considered as organization of knowledge in respect to certain demands and principles. Such demands and principles can vary with the actual knowledge we have. On the basis of what we know we may see differences that we would not otherwise notice. In some respects chemistry and physics are very different types of knowledge, and in other respects they are just parts of the same body of knowledge. Forms of knowledge will then be more basic, in that we will always have to distinguish between two different forms of knowledge, even if they both would have to be involved in any knowledge situation. It is like the typologies of color that can vary, even if we all agree that there are differences in color. Although there is a clear distinction between the categories color and weight, any object which has weight must also have color, and therefore the two concepts are in certain situations, inseparable. To continue the color–weight analogy it is important that colors can be transformed, but that they cannot be transformed into weight. So we can talk of transforming or translating types of knowledge, that we have the same knowledge exemplified or manifested in different ways, but we cannot transform or translate different forms of knowledge. This is important in relation to the issue of tacit knowing. If it is a form of knowing, then it is basic and cannot be transformed to another form. If it is a type of knowledge, then it can be transformed. So if knowledge is tacit, in the sense that it is so basic that it has never actually been formulated, then it is a type of knowledge. But if it is tacit in the sense that it is impossible to formulate it, but one can identify it and talk about it, then it is a form of knowledge. It is important to note that the type concept used here is different from the one used when talking about types and tokens. The American flag I actually look at is a token of the flag, and together with the other tokens it is the set of American flags. The type is what all these have in common, that makes them tokens or exemplifications of the American flag. But there are many different flags, and they can be classified in different ways, and thereby we get different types of flags. But the concept of flag can also be used to refer to entities that are very different, flags that cannot fly for instance, and then we can start a discussion about the difference between types of flags and forms of flags in the sense outlined above.

What is interesting in understanding the use of the concept of knowledge is the various types and forms and how they interact or do not interact. If we return to the basic distinction between knowledge as a state of knowing or the content of that state, knowledge as that which is known and knowledge in use, the activity which qualifies a person or organization as having the predication 'knows what they are doing' attached, we may look at the relation between the two. An important relation here is exemplified by causal knowledge. When we know about causal relations, then we are able to use that knowledge in action. If I know that in a situation like S, if A is present then B will occur, then I can use that to produce B by acting so that in such situations A will be the case. Factual knowledge about relations in the world becomes procedural. The essential idea here is that some types of knowledge which provide content are necessary conditions for the occurrence of other types of knowledge, namely knowledge in use. The use of the explicit content form is only possible based on some tacit/implicit knowledge. So it is the interplay between various forms and types of knowledge which is interesting. If we look at some of the distinctions that we have encountered so far, we have made the distinction between knowledge in the sense of knowing facts, and knowledge in use. We have also looked at the distinction between knowing that and knowing how, which is in many ways similar. In addition, there is knowing in the sense of knowing why, typically causal knowledge. In actual creation and use of knowledge we also have to have immediate perceptual knowledge, and knowledge of the meaning and use of words which of course involves both knowing facts (the meaning of words and sentences) and knowledge in use, actually being able to talk and understand. A famous distinction is one made by Bertrand Russell between knowledge by acquaintance and knowledge by description (Russell, 1912). I know how red looks, or that this is a dog, or this is Peter in a different way from the way I know that force is mass times acceleration; although I may have knowledge by acquaintance of phenomena explained by such knowledge (such as the actual experience of the acceleration of a heavy object).

An important form of knowledge creation and knowledge use is connected to the advanced industrial society. Here knowledge is important in a new form as disciplinary scientific knowledge produced and available in the context of a well-developed social institution: the research system. The knowledge is available in the disciplines that are themselves a typology of knowledge. Whether there are also several forms or only one, scientific knowledge has been discussed intensively in the philosophy of science. The social use of scientific knowledge is based on the idea of using knowledge in problem-solving of problems that are not themselves intradisciplinary. This is then done by focusing on knowledge which has a causal and procedural form. One can use a non-causal natural law; such as 'always A = f (B)' to create the situation B and then get the result A. We can say that we interpret the law causally through our practice. This is how we use natural science in engineering, use economics in business and use physiology in medicine. We can with one expression 'apply' the knowledge and thereby make it useful. The relation between research and the use of knowledge in the economy is based on such relationships. So is the development of science and research-based professions. High-tech industry, health-care, the business system based on knowledge of psychology, organization and economics are results. Value creation in organizations is understood as the increasingly intensive use of scientific knowledge. The nexus between knowledge of facts and knowledge in use is primarily an instrumental one. In the development of societies and economies where the exchange of knowledge and information is as important, if not more important as the exchange of goods, the variety of forms of knowledge that are involved rise and their role in the various social activities change. What we have seen in the case of the emergence of knowledge management has been the recognition that tacit knowledge is important in a more significant way than expected. It has always been the case that we 'know more than we can tell' as phrased by Polanyi (1966), but that this is of central importance in organizations only became explicit after several experiences in the 1980s and 1990s. That understanding one's customers is important in business has always been the case, but only lately has this been explicitly recognized as an important point which one needs to take into account when organizing and managing innovation and development. Knowledge and learning has always been central to a society, but what has changed is the way the knowledge about such processes has been given a role. Knowing about the forms of knowing involved in practices is central to the performance in a knowledge economy.

NOTES

1 For accounts see 'Causal processes' and 'The metaphysics of causation' (Schaffer, 2008; Dowe, 2008).
2 For an early and influential presentation see the work of B.-Å. Lundvall, such as 'From the economics of knowledge to the learning economy'. Lundvall was one of the first economists to analyse the knowledge concept in economic perspective and see knowledge in use as learning.
3 B. Godin and the Canadian INRS have worked on the history of the concepts of the information society and the knowledge economy through especially an analysis of the statistics connected to these notions. See for example Godin (2006, 2008, 2009).
4 Marx: 'The product of mental labor—science—always stands far below its value; because the labor-time needed to reproduce it has no relation at all to the labor-time required for its original production. For example, a schoolboy can learn the binominal theorem in an hour' (Marx, 1861, Addendum 1, 'Hobbes on Labour, on Value and on the Economic Role of Science').
5 See Barber, (1967, Chapter 6) on Alfred Marshall.
6 For a still interesting discussion of explanations of technical development see Elster (1983), and for a typical historical account of the relation between economic and scientific development see Landes (1998).
7 Incorrect information or misinformation is also in a certain sense information. You think misinformation is information if you do not know that it is incorrect.
8 For discussions see Collin (2011), Zammito (2004) and Hacking (1999).
9 Rolf (1991); Josefson (1987, 1987) on tacit knowledge and the professions.
10 http://www.consilium.europa.eu/uedocs/cms_data/docs/pressdata/en/ec/00100-r1.en0.htm.

REFERENCES

Appleby, J. (2010), *The Relentless Revolution – A History of Capitalism*, New York: W.W. Norton.

Backhaus, J. (1993), 'The university as an economic institution: the political economy of the Althoff system', *Journal of Economic Studies*, **20** (4/5), 8–29.

Barber, W. (1967), *A History of Economic Thought*, London: Penguin Books.
Barth, F. (2002), 'The anthropology of knowledge', *Current Anthropology*, **43** (1), 1–11.
Campbell, J. and O. Pedersen (2011), 'Knowledge regimes and comparative political economy', in Daniel Béland and Robert Henry Cox (eds), *Ideas and Politics in Social Science Research*, New York: Oxford University Press, pp. 167–90.
Coase, R. (1974), 'The lighthouse in economics', *Journal of Law and Economics*, **17**, 357–76.
Collin, F. (2011), *Science Studies as Naturalized Philosophy*, Dordrecht: Springer.
Collins, H. and R. Evans (2007), *Rethinking Expertise*, Chicago: University of Chicago Press.
Cortright, J. (2001), 'New growth theory, technology and learning: a practitioners guide', *Reviews of Economic Development Literature & Practice*, USA: US Economic Development Administration.
Davis, Bret W. (2010), 'The Kyoto school', in Edward N. Zalta (ed.), *The Stanford Encyclopedia of Philosophy*, Summer Edition, available at http://plato.stanford.edu/archives/sum2010/entries/kyoto-school/. Last accessed 9 June 2011.
Dewey, J. (1938), *Logic – The Theory of Inquiry*, New York: Henry Holt & Co.
Dowe, Phil (2008), 'Causal processes', in Edward N. Zalta (ed.), *The Stanford Encyclopedia of Philosophy*, Fall Edition, available at http://plato.stanford.edu/archives/fall2008/entries/causation-process/. Last accessed 9 June 2011.
Dretske, F. (1991), *Knowledge and the Flow of Information*, Cambridge, MA: MIT Press.
Elster, J. (1983), *Explaining Technical Change*, Cambridge: Cambridge University Press.
Godin, B. (2006), 'The knowledge-based economy: conceptual framework or buzzword?', *Journal of Technology Transfer*, **31**, 17–30.
Godin, B. (2008), 'The knowledge economy: Fritz Machlup's construction of a synthetic concept', Project on the History and Sociology of STI Statistics, Working Paper no. 37, p. 33. Published in: The Capitalization of Knowledge. A Triple Helix of University-Industry-Government. 2010.
Godin, B. (2009), *Making Science, Technology and Innovation Policy: Conceptual Frameworks as Narratives*, Montreal: Centre Urbanisation Culture Societe, Institut national de la recherche scientifique.
Greenough, P. and D. Pritchard (eds) (2009), *Williamson on Knowledge*, Oxford: Oxford University Press.
Hacking, I. (1999), *The Social Construction of What?*, Cambridge, MA: Harvard University Press.
Hounshell, D. (1984), *From the American System to Mass Production – 1800–1932: The Development of Manufacturing Technology in the United States*, Baltimore, MD: Johns Hopkins University Press.
Jensen, H.S. (2008), 'Management and learning in the knowledge economy', *Journal of Regional Analysis and Policy*, **38**, 130–37.
Josefson, I. (1987), 'Knowledge and experience', *Applied Artificial Intelligence*, **1** (2), 173–80.
Josefson, I. (1987), 'The nurse as an engineer, the theory of knowledge in research in the care sector', *AI in Society*, **1** (2), 115–26.
Kratwohl, D.R. (2002), 'A revision of Bloom's taxonomy, an overview', *Theory into Practice*, **41**, 212–18.
Landes, D.S. (1998), *The Wealth and Poverty of Nations – Why Some are so Rich and Others so Poor*, New York: W.W. Norton.

Lundvall, B.-Å (2000), 'From the economics of knowledge to the learning economy', in *Knowledge Management in the Learning Economy*, Paris: OECD.

Lyotard, J.F. (1979), *La Condition Postmoderne: Rapport sur le Savoir*, France: Les Édition de Minuit, France. English edition (1984): *The Post-modern Condition: A Report on Knowledge*, USA: University of Minnesota Press.

Marx, K. (1861), http://www.marxists.org/archive/marx/works/1863/theories-surplus-value/add1.htm.

Mokyr, J. (2002), *The Gifts of Athena: Historical Origins of the Knowledge Economy*, Princeton, NJ: Princeton University Press.

Molander, B. (1996), *Kunskap i Handling*, Göteborg: Bokförlaget Daidalos.

Nonaka, I. (1991), The knowledge creating company, *Harward Business Review*, **69** (November–December), 96–104.

OECD (Organisation for Economic Co-operation and Development) (1996), *The Knowledge Based Economy*, Paris, OECD.

Peirce, C.S. (1877), 'How to make our ideas clear', in N. Houser and C. Kloesel (eds), *The Essential Peirce – Selected Philosophical Writing Vol. 1*, Indianapolis: Indiana University Press, pp. 124–42.

Penrose, E. (1959), *The Theory of the Growth of the Firm*, New York: John Wiley and Sons.

Petticrew, M. and H. Roberts (2006), *Systematic Reviews in the Social Sciences*, Oxford: Blackwell Publishing.

Pintrich, P.R. (2002), 'The role of metacognitive knowledge in learning, teaching and assessing', *Theory into Practice*, **41** (4), 219–25.

Polanyi, M. (1958), *Personal Knowledge: Towards a Post-Critical Philosophy*, Chicago: University of Chicago Press.

Polanyi, M. (1966), *The Tacit Dimension*, Chicago: The University of Chicago Press.

Popper, K. (1972), *Objective Knowledge – an Evolutionary Approach*, Oxford: Oxford University Press.

Porat, M.U. and M.R. Rubin (1977), *The Information Economy*, Washington: Department of Commerce.

Prusak, L. (2001), 'Where did knowledge management come from?', *IBM Systems Journal*, **40**, 1002–7.

Rescher, N. (2009), *Ignorance – On the Wider Implications of Deficient Knowledge*, Pittsburgh, PA: University of Pittsburgh Press.

Rolf, B. (1991), *Profession, Tradition och Tyst Kunskap*, Lund: Nya Doxa.

Romer, P. (1990), 'Endogenous technological change', *The Journal of Political Economy*, **98**, 71–102.

Russell, B. (1912), *The Problems of Philosophy*, London: Williams and Norgate.

Ryle, G. (1949), *The Concept of Mind*, London: Hutchinson.

Schaffer, Jonathan (2008), 'The metaphysics of causation', in Edward N. Zalta (ed.), *The Stanford Encyclopedia of Philosophy*, Fall Edition, available at http://plato.stanford.edu/archives/fall2008/entries/causation-metaphysics/. Last accessed 9 June 2011.

Solow, R.S. (1957), 'Technical change and the aggregate production function', *Review of Economics and Statistics*, **39**, 312–20.

Steup, Matthias (2008), 'The analysis of knowledge', in Edward N. Zalta (ed.), *The Stanford Encyclopedia of Philosophy*, Fall Edition, available at http://plato.stanford.edu/archives/fall2008/entries/knowledge-analysis/. Last accessed 9 June 2011.

Stewart, T. (1997), *Intellectual Capital: The New Wealth of Organization*, New York: Currency Doubleday.

Taylor, F. (1911), *The Principles of Scientific Management*, London: Harper & Brothers.

Tsoukas, H. (2003), '*Do we really understand tacit knowledge?*' in M. Easterby-Smith and M.A. Lyles (eds), *Handbook of Organizational Learning and Knowledge Management*, Malden: Blackwell, pp. 104–21.

Unger, P. (1975/2002), *Ignorance – a Case for Skepticism*, New York: Oxford University Press.

Wallace, D. (2007), *Knowledge Management – Historical and Cross-Disciplinary Themes (Libraries Unlimited Knowledge Management Series)*, USA: Libraries Unlimited.

Warsh, D. (2006), *Knowledge and the Wealth of Nations*, London/New York: W.W. Norton.

Wittgenstein, L. (1953), *Philosophical Investigations*, New York: Blackwell Publishing.

Zammito, J.H. (2004), *A Nice Derangement of Epistemes*, Chicago: Chicago University Press.

3. Growth of the knowledge-based economy in a two-person non-cooperative game

Hanas A. Cader and John C. Leatherman

INTRODUCTION

The last decade of the twentieth century witnessed remarkable growth in global output, technology, knowledge generation and transfer, transportation, and the proliferation of information. These changes were visible at all organizational levels from the global to the individual firm. These changes stem from one of the fundamental economic principles of our time originally articulated in the late nineteenth century. Alfred Marshall recognized that 'knowledge is our most powerful engine of production' (Marshall, 1890, p. 115). After 100 years, nearly all nations have embraced the concept that knowledge is an important factor of production. In Romer's (1986, 1990) view, knowledge is the third important factor of production, while Peter Drucker (1999) argued that knowledge is the most valuable asset of a twenty-first century firm. As a result, firms have become 'knowledge-intensive organizations' (Florida, 1995).

The significance of knowledge in production has been recognized beyond the boundaries of individual firms. As knowledge-intensive firms find, generate, use and disseminate knowledge, there are proximate, peripheral and remote beneficiaries. The degree of benefit depends on the proximity among knowledge-intensive firms. An increasing number of knowledge-seeking firms have clustered into knowledge-intensive geographic regions. These knowledge-specialized regions have been called the 'learning region' (Florida, 1995; Boekema et al., 2000) or the 'knowledge-based city' (Simmie and Lever, 2002). Alternatively, knowledge clusters can exist in the virtual world, given rise by common needs and interests and enabled through telecommunications and other information and communication technologies (Döring and Schnellenback, 2006; Russ and Jones, 2011).

The evolution and growth of knowledge-intensive cities was not without consequence. The clustering of firms in regions increases the competition for knowledge generation and use in value creation. Simmie (2002) called these competitive, knowledge-intensive regions the 'competitive city'. To compete and survive, firms are forced to innovate. In Porter's (1990) terminology, these regions form an 'innovative cluster'. Employees of firms in every economic sector became the 'collective agents of innovation' (Florida, 1995, p. 529).

The basic building blocks of knowledge-intensive regions are the individuals comprising the workforce. Assessment of a knowledge-intensive region requires measuring the level of knowledge of individuals within the workforce. The difficulty with that is that there are no standard criteria by which to measure the knowledge level of individuals. One could argue that educational attainment would be a measure of knowledge assets. By itself, educational attainment would be an imperfect proxy because it fails to capture significant knowledge components such as experiential learning and the context of the workplace environment. Further, linking educational attainment, experience, and so on, to productivity remains a challenge.

Machlup (1962) suggested that the knowledge intensity of a region could be measured by assessing knowledge assets, while Eliasson et al. (1990) said it could be determined by the value of labor in knowledge-intensive sectors. One recent study suggested that firms could manage tacit and explicit knowledge through the environmental knowledge circulation process (EKCP) to improve corporate performance (Huang and Shih, 2010) These measures are very much subjective in nature.

In the literature, we find three measures of the knowledge intensity of a region that fulfill objectivity criteria reasonably well: evaluating the labor qualifications of workers (Burton-Jones, 1999); determining the degree to which industry sectors use advanced machines and management practices (McGranahan and Beale, 2002); and gauging the relative proportion of an industry's labor force comprising certain managerial and professional/technical occupations (Beck, 1992). More recently, Raspe and van Oort (2008) used a hybrid approach combining multiple indicators of knowledge intensity (e.g. education, technology use, occupational structure, etc) to characterize regions. These criteria can be applied to measure the knowledge intensity of a regional or national economy. Given the increasing importance of the overall knowledge level found within the economy due to global competitiveness, we utilize these concepts in characterizing the 'new' economy, the 'knowledge' economy, or the 'digital' economy.

Knowledge-intensive workers require higher-order cognitive skills compared to non-knowledge workers, and can be identified by their occupational titles (Cader, 2008). Knowledge-intensive occupations include managerial, professional and technical occupations (Beck, 1992). One can assess the knowledge-based economy by measuring the relative proportion of managerial, professional and technical occupations in the labor force (Wojan, 2000; Beck, 1992; Cader, 2008). Alternatively, Raspe and Van Oort (2008) used the average educational level of the workforce as an indicator of knowledge economy. Growth of the knowledge-based economy both spurs and depends upon growth of a knowledge-intensive labor force. The cause and effect of growth of the knowledge-based economy versus growth of a knowledge-intensive workforce is of secondary concern to the recognition of the 'virtuous cycle' that is created. Clearly, it has to start with an idea about how a process can be improved or made more efficient in response to a competitive environment. This creates demand for the capacity to implement the innovation. Workers with such capacity, ever responding to a competitive environment, innovate further, creating new economic opportunity and new demands for a capable workforce. Growth of the knowledge-intensive workforce increases the stock of knowledge in the economy which creates new knowledge (knowledge breeds knowledge).

Individuals make occupational choices based on their broadly defined self-interest (remuneration, status, power, aspirations, etc.) and the expected pay-off of occupational achievement. Individuals make strategic choices to secure a given occupation in an often competitive environment. Further, an individual's choices and actions may affect another's behavior. This might be the case, for example, if there is a vacant managerial position in a firm and two employees hope to fill the slot. To enhance their competitive advantage, one of the employees may chose to increase their knowledge through additional education, while the other may stay in the office and work longer hours. Similarly, when making a career choice, individuals consider different occupational characteristics, such as salary level, industry growth, employment prospects and other factors that would collectively provide the highest level of satisfaction. These strategies can be examined using a game theory perspective, where players react to different alternatives available to themselves and others. The important fact is that the end result depends not only on each player's individual actions, but on all players' actions. As individuals pursue their self-interest, they are less likely to cooperate with each other, resulting in a non-cooperative response equilibrium.

The main question we want to analyse in this chapter is whether the natural competition between workers for occupational advancement and the attendant increasing pay-off associated with higher-order occupations will lead to growth in knowledge-based occupations. This is, to our knowledge, the first attempt to analyse growth of the knowledge-based economy using a game theoretic perspective. In this chapter, we present the individual's knowledge-intensive occupational achievement process and the resulting theoretical equilibrium using non-cooperative game theory and examine the growth of knowledge-based industry. The remainder of the chapter is organized into four sections. The next section discusses non-cooperative game theory and an illustration of the outcome of the competition between workers vying for an occupational advancement opportunity. In the following section, we offer empirical evidence of knowledge-based industry growth in the US economy that would seem to validate our predicted outcome. In the final section, we present summary findings and the conclusion of the study.

NON-COOPERATIVE GAME THEORY MODEL

In this section, we discuss the underlying game theory model used to explain the process of knowledge economy growth in the context of self-interested economic agents maximizing their utility. Assume there is a knowledge-based occupation position opening and two non-knowledge-based workers (players) trying to secure the position. In this chapter, we use a non-cooperative game theory model because cooperation between the two players is highly unlikely in pursuit of the position. Further, each player will adopt individual competitive strategies and is unlikely to share information with the other.

Players

Our model begins with the assumption that there are two types of workers in the economy. There are knowledge workers and non-knowledge workers. Knowledge workers possess specific levels of knowledge and skill needed to perform an occupation. In our model, workers in occupational categories of managers, professionals and technical occupations are considered to be knowledge workers. All other occupational categories are considered to be non-knowledge workers. In the game theory model, there are two individuals **i** and **j** with distinct qualifications Q**i** and Q**j**. These two individuals work in a firm as non-knowledge workers and receive a salary of **Wi** and **Wj**, respectively. Assuming that a knowledge-based position becomes available, individuals **i** and **j** compete to move into the knowledge-based occupational opportunity.

The Rules

At the start of the game, both players are assumed to have an equal opportunity to be selected for the knowledge-intensive vacancy, but only one will be chosen. The players have the freedom to employ a variety of strategies to increase the likelihood that they will be the one chosen for the knowledge-based occupation.

Outcomes

Only one of the players can be selected for the knowledge-based occupation and the other will not become a knowledge-based worker. While the one who was not selected can look for a knowledge-based occupation elsewhere, in this case, we assume that individual will continue to be a non-knowledge-based worker. The only other outcome is that neither player is selected for the job.

Pay-offs

Of course, there are multiple incentives for occupational aspiration. For simplicity, we assume that the pay-off is the remuneration the players receive in either being selected or not being selected for the knowledge-based occupation. We assume the knowledge-based worker's salary is sufficiently desirable to motivate the desire to move to the new occupation. The players' pay-off depends on whether they are selected. Player **i**'s and player **j**'s pay-offs and outcomes can be presented in the normal form of the game or game matrix.

The growth of the knowledge-based economy depends on the outcome of the game. Since each individual is likely to attempt to maximize his/her pay-off, each will employ strategies to win the game. In this game, one player being selected does not have a negative impact on the other. The game can be presented in the normal form shown in Table 3.1.

Table 3.1 Pay-off matrix for players i and j

		Player **j**	
		Selected	Not selected
Player **i**	Selected	0 0	m a
	Not selected	b m	C c

The pay-off in each cell's top-left corner is the pay-off for player **i**, and bottom-right for player **j**. If player **i** is selected (player **j** not selected), then player **i**'s pay-off is **m** and player **j**'s pay-off is **a**. If player **j** is selected (player **i** not selected), then player **j**'s pay-off is **m** and player **j**'s pay-off is **b**. If the firm selects neither player, then one or both may choose to leave the firm or continue in their current non-knowledge-based occupation. For simplicity, we do not attempt to model this outcome. Further, neither player is assumed to have a dominant strategy, one which will make one player more likely to be selected for the vacant knowledge-based position. The individual **i**'s or **j**'s selection depends on their pay-off in the case of being selected or not selected, where **m** > **c** > **a** or **b**. The arrows outside the pay-off matrix indicate each player's preferred direction of movement. Player **i** prefers upward movement, moving from not being selected to selected, while player **j** prefers horizontal movement, going from not being selected to selected. The Nash equilibrium depends on the interaction of different pay-offs that the players will receive in the event of being selected or not selected. In this game, there are two Nash equilibriums. The first Nash equilibrium is the probability of player **i** being selected, which can be stated as:

$$p_i = \frac{(m - c)}{(m - c + a)} \tag{1}$$

and the second is player **j** being selected:

$$p_j = \frac{(m - c)}{(m - c + b)} \tag{2}$$

This represents the standard solution from a game theoretic perspective.

MEASURING KNOWLEDGE, OCCUPATIONS AND OCCUPATIONAL PAY-OFFS

Measuring the stock of knowledge in a region or assessing its contribution to economic development is extremely difficult. Moreover, there are no universally accepted criteria by which to measure workers' knowledge (Ramirez and Nembhard, 2004). Shapira et al. (2006, p. 1528) noted that there is not even a proxy to measure knowledge. Many studies have attempted to measure knowledge either at the firm or individual worker level (Drucker, 1999; Pyöriä et al., 2005; Harrison et al., 2004; Nonaka and Takeuchi, 1995; Westeren, 2008). It is possible to develop indicator sets saying something about the knowledge level of individual firms. In most countries, firms are allowed to include knowledge capital in their balance sheets. But it is difficult to specify a more general formulation of a firm's level of knowledge.

Knowledge use or expression does not have time and space relevance as it can be used to add value in any situation. As a result, some recent research characterized knowledge work as firm- and situation-specific (Scott, 2005; Quinn, 2005; Westeren, 2008). This approach is very much in contrast to the argument presented by Shapira et al. (2006) that suggests the measurement of knowledge at the industry scale provides a more detailed comparison in contrast to macro- or micro-level measures. At the firm level, there is a strong correlation between the knowledge level of workers and firm performance (Kubo and Saka, 2002), and the key players in the organization are the knowledge workers

(Wickramasinghe and Ginzberg, 2001). In organizations, knowledge workers often are identified based on the function that he/she performs (Cader, 2008) or based on their nature of work (Scarbrough, 1999). Employee occupational title is one method used to identify a knowledge worker. An employee's occupational title reflects the skill level required to perform an occupation (Cader, 2008). In this paper we used the occupational title of employees to identify knowledge and non-knowledge workers in the economy.

In the USA, occupational titles of workers are annually published by the US Department of Labor in the 'Dictionary of Occupational Titles' that identifies the different level of cognitive skills that are required for a variety of occupations. The General Educational Development-Reasoning Scale (GEDR) is used to measure the skill level of occupations. The GEDR scale is an ordinal ranking of cognitive requirements to perform an occupation ranging from the ability to follow simple instructions to the ability to solve complex problems (Wojan, 2000, p. 598). While ordinary workers (laborers) need/possess a low level of cognitive skills, knowledge workers are required to have a higher level of cognitive skills (Raspe and van Oort, 2008). The required level of cognitive skills increases as the rank of the occupation increases from lower to higher ranks. The GEDR for different occupational classifications is presented in Table 3.2.

DATA

In this study, we used the US National Occupational Employment Statistics (OES) data. The OES data consist of national- and state-level occupational statistics for all industries. The Bureau of Labor Statistics (BLS) of the US Department of Labor publishes these data annually. The OES data provide information on occupational employment and wage rates for wage and salary workers in non-farm establishments. The BLS collects these data through a survey of employers in all industry sectors. According to the BLS publication (2009), all workers are classified into one of over 820 occupations according to their occupational definition. To facilitate classification, occupations are combined to form 23 major groups, 96 minor groups, and 449 broad occupations. Although there are no official criteria by which to group these 23 major occupations as knowledge-based or non-knowledge-based, Cader (2008, p. 124) classified 10 of the 23 major occupational groups as knowledge-based occupations:

1. management occupations;
2. business and financial operations occupations;
3. computer and mathematical occupations;
4. architecture and engineering occupations;

Table 3.2 *Skills requirements distribution (%) by summary occupations unweighted detail occupations from the dictionary of occupational titles*

	General Educational Reasoning Scale					
Occupational category	GEDR 1	GEDR 2	GEDR 3	GEDR 4	GEDR 5	GEDR 6
Managers	0	0	4.54	44.59	46.16	4.71
Professional/ technical	0	0.23	2.80	18.27	42.17	36.51
Sales	0.15	2.25	21.35	64.81	11.42	0
Technical	0.23	7.20	56.55	34.59	1.29	0.11
Precision production	0.99	10.94	29.77	58.02	0.26	0
Operators	5.62	50.75	37.42	5.97	0.21	0.02
Service occupations	2.94	24.88	46.15	23.07	2.94	0
Laborers	38.52	57.32	4.14	0	0	0
Farm, forestry and fisheries	8.08	26.41	27.49	26.68	11.32	0
All occupations	4.65	23.96	25.11	26.49	12.32	7.43

GEDR 1. Apply common sense understanding to carry out simple one- or two-step instructions. Deal with standardized situations with occasional or no variables in or from these situations encountered on the job.

GEDR 2. Apply common sense understanding to carry out detailed but uninvolved written or oral instructions. Deal with problems involving a few concrete variables in or from standardized situations.

GEDR 3. Apply common sense understanding to carry out instructions furnished in written, oral, or diagrammatic form. Deal with problems involving several concrete variables in or from standardized situations

GEDR 4. Apply principles of rational systems* to solve practical problems and deal with a variety of concrete variables in situations where only limited standardization exists. Interpret a variety of instructions furnished in written, oral, diagrammatic, or schedule form. *Examples of rational systems are: bookkeeping, internal combustion engines, electric wiring systems, house building, farm management, and navigation.

GEDR 5. Apply principles of logical or scientific thinking to define problems, collect data, establish facts, and draw valid conclusions. Interpret an extensive variety of technical instructions in mathematical or diagrammatic form. Deal with several abstract and concrete variables.

GEDR 6. Apply principles of logical or scientific thinking to a wide range of intellectual and practical problems. Deal with nonverbal symbolism (formulas, scientific equations, graphs, musical notes, etc.) in its most difficult phases. Deal with a variety of abstract and concrete variables. Apprehend the most abstruse classes of concepts.

Source: US Department of Labor, 1991, 1994

5. legal occupations;
6. arts, design, entertainment, sports and media occupations;
7. healthcare practitioners and technical occupations;
8. life, physical and social science occupations;
9. education, training, and library occupations;
10. healthcare practitioners and technical occupations.

From the 23 major occupations, we have chosen one knowledge-based occupation (1 below) and four non-knowledge-based occupations (2–5 below) for the purpose of this study. They are:

1. life, physical, and social science occupations (SOC code = 19-0000);
2. protective service occupations (SOC code = 33-0000);
3. sales and related occupations (SOC code = 41-0000);
4. office and administrative support occupations (SOC code = 43-0000);
5. production occupations (SOC code = 51-0000).

These five occupational groups were selected to represent a graduated GEDR scale. That is, we assume that the knowledge and skills necessary to perform each successive occupation is incrementally and gradually more complex. This assumption was in line with the recent empirical findings of Cordes (2011). For example, a production-related occupation requires more of the GEDR level 2 skills while a service-related occupation requires more GEDR level 4 skills. For simplicity, we also presume that each successive occupation is valued progressively more. In the real world, of course, we observe lower-order occupations that may pay financial premiums for risk or difficulty, but this would not change our overall conclusions.

We used the mean annual salary data of these four major non-knowledge-based occupations to estimate the probability of a non-knowledge-based worker moving to a knowledge-based occupation in the states of Massachusetts and Rhode Island. While these are contiguous states, there is a considerable variation in the characteristics of their economies. We use 2005 and 2008 data to estimate the probabilities. The SOC codes for 23 major occupations, their descriptions, annual employment and mean annual salary for Massachusetts and Rhode Island are found in Appendix 3.1. The OES data are available from 1997 to 2008 and can be downloaded from the BLS website. These two years were selected to allow a three-year gap between successive time periods. It would take approximately three years for a college sophomore to look at the current OES data, make a choice regarding his/her college major, and graduate at the end of the third year. This individual will enter the job market to find a job related to his/her major (knowledge-based or non-knowledge-based). The data used in the analysis are presented in Table 3.3.

Table 3.3 Occupations and mean annual salaries in 2005 and 2008

		Massachusetts		Rhode Island	
Occupations	SOC Code	2005	2008	2005	2008
Life, physical, and social science occupations	19-0000	$65 320	$73 490	$57 800	$65 350
Protective service occupations	33-0000	$39 740	$44 520	$38 320	$42 540
Sales and related occupations	41-0000	$38 110	$41 750	$30 900	$35 240
Office and administrative support occupations	43-0000	$34 470	$37 140	$31 030	$33 430
Production occupations	51-0000	$33 290	$34 990	$29 470	$30 950

Source: US Department of Labor, BLS, 2009

RESULTS AND DISCUSSION

The Nash equilibrium probabilities from models (1) and (2) were estimated using the data in Table 3.3. Following from Table 3.1, there are four pay-offs, and in Table 3.3, there are salaries for five occupations. We combine Table 3.1 and Table 3.3 in the following manner. At the beginning, players **i** and **j** belong to a non-knowledge-based occupation. According to Table 3.3, both players belong to production occupations. If player **i** or **j** is selected to be a knowledge-based worker (the pay-off is **m**), then their occupation would be life, physical and social science occupations (annual salary in MA in 2008 is $73 490). If player **i** and **j** are not selected, then both would be in the protective services occupational category and their annual salary would be $44 520 in MA in 2008. If player **i** is not selected while **j** is selected to be a knowledge-based worker (life, physical and social science occupation), then player **i**'s new occupation would be a sales-related occupation. Conversely, if player **j** is not selected while **i** is selected to be a knowledge-based worker (life, physical and social science occupations), then player **j**'s new occupation would be office and administrative support. If we replace the pay-offs in Table 3.1 with the mean annual salary in 2008 for Massachusetts, then we derive the following Table 3.4, assuming that other benefits are equal for all the occupations.

Table 3.4 Mean annual salary (in US dollars) pay-off matrix for players i and j for MA in 2008

		Player **j** Selected	Player **j** Not selected
Player **i**	Selected	\$0 \$0	\$73 490 \$37 140
	Not selected	\$41 750 \$73 490	\$44 520 \$44 520

We can replace the data in Table 3.4 for MA in the year 2005, RI in 2005 and 2008, and estimate the probabilities for player **i** and player **j** presented in Table 3.5.

The results show that in 2008, both players have higher probabilities of becoming a knowledge worker than in 2005. This result appears to show a natural growth of knowledge-based occupations. Over time, the workers have an opportunity to acquire new knowledge (learning by doing), develop new skills, and learn through interaction with peers and superiors. Over a given time horizon, the players can identify, plan and implement their strategies for advancement. In time, groups within the workforce can form and implement initiatives to advance their knowledge skills. In related research, Davenport (2005) suggested that knowledge workers were likely to collaborate and congregate with like workers. It appears to be true for potential knowledge-based workers as well. Further, we deliberately chose the three-year time gap to allow occupational pay-off information to be made available to a college sophomore, who may then choose an occupation more likely to provide a better return for the education. Knowledge-based occupations are relatively well paid and, indeed, it seems to be working in the market.

Table 3.5 Estimated probabilities for players i and j to be selected to become a knowledge-based worker

	Year	Probability for player **i**	Probability for player **j**
Massachusetts	2005	0.426	0.402
	2008	0.438	0.410
Rhode Island	2005	0.386	0.387
	2008	0.406	0.393

The results also indicate that players **i** and **j** have higher probabilities of being chosen as a knowledge-based worker in Massachusetts than in Rhode Island in both years. The underlying socioeconomic characteristics and physical infrastructure of the states could play an important role in facilitating and attracting knowledge-based occupations and industries. Massachusetts is the home of many top tier universities, including Harvard and MIT. Player **i** is more likely to be selected to be a knowledge-based worker than player **j** in Massachusetts. The outcome of the game very much depends on the pay-offs the players receive. The pay-off for player **j** if not selected is less than if player **i** is not selected. Since player **j**'s pay-off is less (is not selected), player **i** would develop strategies to avoid being in that position. This naturally helps player **i**'s chances of being selected for the knowledge-based occupation. On the other hand, in Rhode Island, players **i** and **j** have almost an equal probability of being selected for the knowledge-based occupation. This is primarily due to the pay-offs of both players if both are not selected. The pay-off is nearly equal for player **i** and player **j**.

It is equally true, both empirically and logically, that a non-knowledge worker may have an incentive to move to a knowledge-based occupation even if the salary or other benefits are lower in knowledge-based jobs for reasons other than pecuniary gain. Based on our model, however, the expected probability of that individual choosing the knowledge-based occupation is zero. The probability of moving to a knowledge-based occupation largely depends on the relative difference of the salaries or other perceived non-financial benefits.

CONCLUSION

In this chapter, we studied the potential growth of the knowledge-based economy using a game theory approach. We hypothesized that competition between workers for career advancement would lead to a natural growth of knowledge-based occupations. Both our theoretical and empirical findings confirm the notion. We offer a novel perspective providing a theoretical basis tentatively confirmed by empirical analysis to better understand the seeming 'natural' growth of the knowledge-based economy so ubiquitous around the globe. We suggest the application of game theory as a means by which future investigation can probe important economic phenomena that impinge on the welfare of nations. The challenge for future investigation will be to increase the complexity of analysis to reflect the interactions and interdependencies known to exist among economic agents and institutions. A game theoretic perspective may yield new insight into the economic welfare of nations.

The pay-offs the players receive in being selected to enter the knowledge-based economy determines the probability of each players' selection. The larger the difference in the pay-off of being selected versus not, the higher the probability of movement toward a knowledge-based occupation. Our model shows that the key factor determining the growth of the knowledge-based economy is occupational wage levels. Occupational wages should reflect the marginal value of labor and should be determined by market forces. The knowledge worker possesses the knowledge likely to add value to the production process and thereby be rewarded with higher wages. Since the knowledge-based economy is dynamic, a worker's career opportunity often leads to changing positions. Workers in their current employment positions are likely to compete for better positions. As Hirsch (1977, p. 5) suggested, 'if everyone stands on tiptoe, no one sees better'. The competition is natural. Hence, the model predicts that if knowledge-based occupations are additionally compensated, the knowledge economy will continue to grow. Further, a higher wage for knowledge-based occupations would attract more potential knowledge-based workers into the labor market. The empirical results from Massachusetts and Rhode Island support our game theory findings. Further, we also estimated the percentage of total knowledge workers in ten knowledge-based occupations for MA and RI in 2005 and 2008. The percentage of knowledge-based workers in the years 2005 and 2008 for MA are 38.2 percent and 39.5 percent, respectively, while for RI it was only 30.2 percent and 30.6 percent, respectively.

The implications of our findings are several. The extent to which national economies will grow and prosper in an environment characterized by freer trade and mobility will depend on the institutional rules and structures that are formulated and enforced. Economies characterized by insular industries and workforce protection are not likely to foster the competitive environment whereby either firms or individual workers aspire to innovate with the concomitant rewards. Obviously, this calls for incentive structures whereby knowledge-based advances are appropriately emphasized and rewarded. Clearly, in economies that stress protection of industries, firms and workers from competitive forces, we would predict slower growth of knowledge-based economic opportunity.

The lesson for more market-oriented economies would be one where clear incentive and reward structures are in place to foster innovation and the 'virtuous cycle' of economic and labor force advancement. To that end, public policy might play a role even in advanced economies to incentivize firm and industry behavior in ways that reward innovation over cost-cutting and quarterly profit motivations.

Our results suggest a 'natural' growth of the knowledge economy under conditions of a competitive marketplace with appropriate incentive structures. One logical policy approach to fostering such growth for individual workers would be to facilitate information access about occupational opportunities and incentive structures. In the case of our hypothetical college sophomore, facilitating an understanding of the attendant financial and non-financial rewards associated with alternative career choices, should lead to relatively higher-order occupational aspiration. Similarly, ensuring that workers have information about occupational advancement opportunities and the training and transitional assistance to pursue them would be another way to foster knowledge economy growth.

In summary, we offered a novel approach to understanding change in the knowledge economy. Our use of game theory provides a logical and intuitively appealing perspective, and one that was validated by empirical evidence. Successively more complex applications of a game theory approach may lead to other useful insights about the nature of the knowledge-based economy and what can be done to facilitate new economic growth opportunities.

REFERENCES

Beck, N. (1992), *Shifting Gears: Thriving in the New Economy*, Toronto: Harper Ed.

Boekema, F., K. Morgan, S. Bakkers and R. Rutten (2000), 'Introduction to learning regions: a new issue for analysis', in F. Boekema, K. Morgan, S. Bakkers and R. Rutten (eds), *Knowledge, Innovation and Economic Growth: The Theory and Practice of Learning Regions*, Cheltenham, UK and Northampton, MA, USA: Edward Elgar Publishing, pp. 3–16.

Bureau of Labor Statistics (BLS) (2009), Standard Occupational Classification (SOC), Washington, DC: BLS.

Burton-Jones, A. (1999), *Knowledge Capitalism*, Oxford: Oxford University Press.

Cader, H.A. (2008), 'The evolution of knowledge economy', *Journal of Regional Analysis & Policy*, **38** (2), 117–29.

Cordes, A. (2011), 'What drives skill-biased regional employment growth in West Germany?', in K.I. Westeren (ed.), *Foundations of the Knowledge Economy*, Cheltenham, UK and Northampton, MA, USA: Edward Elgar Publishing.

Davenport, T.H. (2005), *Thinking for a Living: How to Get Better Performance and Results for Knowledge Workers*, Boston: Harvard Business School Press.

Döring, T. and J. Schnellenback (2006), 'What do we know about geographical knowledge spillovers and regional growth? A survey of the literature', *Regional Studies*, **40** (3), 375–95.

Drucker, P.F. (1999), *Management Challenges for the 21st Century*, Butterworth-Heinemann: Oxford.

Eliasson, G., S. Folster, T. Lindberg, T. Pousette and E. Taymaz (1990), *The Knowledge Based Information Economy*, Stockholm: The Industrial Institute for Economic and Social Research.

Florida, R. (1995), 'Regional creative destruction: production organization, globalization, and the economic transformation of the Industrial Midwest', *Economic Geography*, **72** (2), 315–35.

Harrison, A., P. Wheeler and C. Whitehead (2004), *The Distributed Workplace*, London: Routledge.

Hirsch, F. (1977), *Social Limits to Growth*, London: Routledge and Kegan Paul.

Huang, P. and P. Shih (2010), 'The impact of industrial knowledge management and environmental strategy on corporate performance of ISO-14000 companies in Taiwan: The application of structural equation modeling', *African Journal of Business Management*, **4** (1), 021–030.

Kubo, I. and A. Saka (2002), 'An inquiry into the motivations of knowledge workers in the Japanese financial industry', *Journal of Knowledge Management*, **6** (3), 262–71.

Machlup, F. (1962), *The Production and Distribution of Knowledge in the United States*, Princeton, NJ: Princeton University Press.

Marshall, A. (1890), *Principles of Economics*, 8th edn, London: Macmillan and Co. Ltd.

McGranahan, D.A. and C.L. Beale (2002), 'Understanding rural population loss', *Rural America*, **17** (Winter), 1–10.

Nonaka, I. and H. Takeuchi (1995), *The Knowledge-creating Company: How Japanese Companies Create the Dynamics of Innovation*, New York: Oxford University Press.

Porter, M.E. (1990), *The Competitive Advantage of Nations*, New York: Free Press.

Pyöriä, P., H. Melin and R. Blom (2005), *Knowledge Workers in the Information Society. Evidence from Finland*, Tampere: Tampere University Press.

Quinn, R.W. (2005), 'Flow in knowledge work: high performance experience in the design of national security technology', *Administrative Science Quarterly*, **50,** 610–41.

Ramírez, Y. W. and D.A. Nembhard (2004), 'Measuring knowledge worker productivity: A taxonomy', *Journal of Intellectual Capital*, **5** (4), 602–928.

Raspe, O. and F. van Oort (2008), 'Localized knowledge externalities in regional economic development and firm growth', *Journal of Regional Analysis & Policy*, **38** (2), 100–16.

Romer, P.M. (1986), 'Increasing returns and long-run growth', *Journal of Political Economy*, **94** (5), 1002–37.

Romer, P.M. (1990), 'Endogenous technological change', *Journal of Political Economy*, **98** (5), 71–102.

Russ, M. and J.K. Jones (2011), 'International virtual industry clusters and SMEs: early process policy recommendations', in Westeren, Knut Ingar (ed.), *Foundations of the Knowledge Economy*, Cheltenham, UK and Northampton, MA, USA: Edward Elgar Publishing,

Scarbrough, H. (1999), 'Knowledge as work: conflicts in the management of knowledge workers', *Technology Analysis & Strategic Management*, **11** (1), 5–16.

Scott P.B. (2005), 'Knowledge workers: social, task and semantic network analysis', *Corporate Communications: An International Journal*, **10** (3), 257–77.

Shapira, P., J. Youtie, K. Yogeesvaran and J. Zakiah (2006), 'Knowledge economy measurement: methods, results and insights from the Malaysian knowledge content study', *Research Policy*, **35** (10), 1522–37.

Simmie, J. (2002), 'Trading places: competitive cities in the global economy', *European Planning Studies*, **10** (2), 201–14.

Simmie, J. and W.F. Lever (2002), 'Introduction: The knowledge-based city', *Urban Studies*, **39** (5–6), 885–57.

US Department of Labor (1991), *Dictionary of Occupational Titles*, Employment and Training Administration, US Employment Service, Washington, DC: US Department of Labor.

US Department of Labor (1994), *Dictionary of Occupational Titles*, Employment and Training Administration, US Employment Service, Washington, DC: US Department of Labor.

Westeren, K.I. (2008), 'How to define and measure knowledge for the analysis of competitiveness', *Journal of Regional Analysis & Policy*, **38** (2), 138–44.

Wickramasinghe, N. and M.J. Ginzberg (2001), 'Integrating knowledge workers and the organization: The role of IT', *International Journal of Health Care Quality Assurance*, **14** (6), 245–53.

Wojan, T. (2000), 'The composition of rural employment growth in the "New Economy"', *American Journal of Agricultural Economics*, **82** (August), 594–605.

APPENDIX 3.1

Occupational statistics for Massachusetts and Rhode Island, 2005 and 2008

State	Occupational code	Occupational title	Employment 2005	Employment 2008	Annual salary 2005	Annual salary 2008
MA	00-0000	All occupations	3 144 540	3 234 860	45 970	50 990
MA	11-0000	Management occupations	197 650	185 940	99 600	114 060
MA	13-0000	Business and financial operations occupations	157 940	171 590	67 030	73 640
MA	15-0000	Computer and mathematical occupations	110 190	121 680	77 450	86 760
MA	17-0000	Architecture and engineering occupations	73 800	77 030	70 620	79 630
MA	19-0000	Life, physical and social science occupations	42 530	53 900	65 320	73 490
MA	21-0000	Community and social services occupations	56 640	65 800	40 060	42 760
MA	23-0000	Legal occupations	25 100	25 550	91 850	100 360
MA	25-0000	Education, training and library occupations	208 660	215 880	50 060	54 300
MA	27-0000	Arts, design, entertainment, sports and media occupations	45 650	51 830	49 280	55 150
MA	29-0000	Healthcare practitioners and technical occupations	193 290	208 500	64 440	77 300
MA	31-0000	Healthcare support occupations	90 160	100 040	27 940	31 060

MA	33-0000	Protective service occupations	72 310	72 220	39 740	44 520
MA	35-0000	Food preparation and serving related occupations	258 980	267 360	21 750	24 280
MA	37-0000	Building and grounds cleaning and maintenance occupations	105 390	104 090	26 560	29 210
MA	39-0000	Personal care and service occupations	*	82 680	26 470	28 450
MA	41-0000	Sales and related occupations	326 420	334 150	38 110	41 750
MA	43-0000	Office and administrative support occupations	538 020	547 620	34 470	37 140
MA	45-0000	Farming, fishing and forestry occupations	2 660	2 700	25 010	26 760
MA	47-0000	Construction and extraction occupations	115 220	108 160	48 160	52 690
MA	49-0000	Installation, maintenance and repair occupations	100 580	103 220	43 930	46 750
MA	51-0000	Production occupations	184 080	173 110	33 290	34 990
MA	53-0000	Transportation and material moving occupations	167 630	161 800	31 240	33 670
RI	00-0000	All occupations	479 810	478 420	38 980	43 410
RI	11-0000	Management occupations	23 230	21 450	89 370	105 220
RI	13-0000	Business and financial operations occupations	19 620	19 930	58 960	62 890

State	Occupational code	Occupational title	Employment 2005	Employment 2008	Annual salary 2005	Annual salary 2008
RI	15-0000	Computer and mathematical occupations	9 750	10 270	64 890	74 570
RI	17-0000	Architecture and engineering occupations	8 040	7 890	67 000	74 210
RI	19-0000	Life, physical and social science occupations	4 080	3 500	57 800	65 350
RI	21-0000	Community and social services occupations	9 780	10 060	38 480	43 180
RI	23-0000	Legal occupations	3 180	3 330	69 380	90 030
RI	25-0000	Education, training and library occupations	32 730	32 780	48 010	54 820
RI	27-0000	Arts, design, entertainment, sports and media occupations	6 510	6 840	43 530	48 720
RI	29-0000	Healthcare practitioners and technical occupations	28 680	30 380	61 650	71 830
RI	31-0000	Healthcare support occupations	17 010	19 420	26 400	29 010
RI	33-0000	Protective service occupations	10 960	10 850	38 320	42 540
RI	35-0000	Food preparation and serving related occupations	44 770	47 340	19 670	21 530
RI	37-0000	Building and grounds cleaning and maintenance occupations	15 500	16 060	24 670	27 090

RI	39-0000	Personal care and service occupations	12 220	12 690	23 690	24 750
RI	41-0000	Sales and related occupations	45 580	44 170	30 900	35 240
RI	43-0000	Office and administrative support occupations	82 200	85 820	31 030	33 430
RI	45-0000	Farming, fishing and forestry occupations	430	*	22 410	23 230
RI	47-0000	Construction and extraction occupations	21 160	18 910	41 220	46 840
RI	49-0000	Installation, maintenance and repair occupations	16 000	16 180	38 590	41 720
RI	51-0000	Production occupations	39 410	35 330	29 470	30 950
RI	53-0000	Transportation and material moving occupations	28 970	24 940	27 770	31 010

* Employment estimate not available.

PART II

Innovation

4. Innovation: from Schumpeter to the knowledge economy

Knut Ingar Westeren

INTRODUCTION

Innovation as an important part of social development has been studied and debated for more than a hundred years. It is not easy to give a comprehensive view of the debate since there have been many different points of departure. The first purpose of this chapter is to discuss some highlights from the conceptual discussions about innovation. One reason for doing this is to investigate if we can find some lines of thought that are relevant to one of today's most challenging discussions regarding the relationship between knowledge and innovation. The other purpose of this chapter is to present new empirical research where we have been investigating innovation and knowledge at the company level. Bringing the conceptual and empirical investigations together forces us to be specific in defining and the use of concepts. It is not enough just to talk about links between knowledge and innovations – the aim in the empirical part of the chapter is to be more specific. How can we define a process innovation in a company and how can knowledge and other factors that contribute to (or prevent) innovations be operationalized? These questions must be discussed before one can analyse relationships between innovations and explaining factors.

HIGHLIGHTS FROM THE CONCEPTUAL AND THEORETICAL DISCUSSION OF INNOVATION

Comments on Schumpeter

A discussion of the innovation concept often starts with Joseph A. Schumpeter, an economist, mathematician and social analyst working in the first half of the 1900s. Schumpeter wanted to address the big questions and analyse the structure and functioning of the capitalist economic system. In Schumpeter's (1926, 1934) early analysis of innovation he identifies a group of people with special abilities and characterizes them as energetic and dynamic. These people constitute a special elite who become 'agents of economic development', constituting the innovative dynamo of the capitalist economy driving the system forward. From this point of departure Schumpeter gives an analysis of the innovation process, from invention to innovation to diffusion.

In brief, one can say that Schumpeter's analysis suggests that innovative shock within the capitalist system provides economic development that proceeds in waves and that these innovative activities contribute to increase society's surplus, including increased real wages. Schumpeter says in his earlier writings that the entrepreneur is the primary source of economic dynamism, while the capitalist market system with prices and competition is the mechanism that allows the entrepreneur to innovate and transfer innovations to social production results.

In the book, *Capitalism, Socialism and Democracy*, Schumpeter (1942/1950, p. 82) presents the concept of creative destruction:

> The opening up of new markets, foreign or domestic, and the organizational development from the craft shop and factory to such concerns as U.S. Steel illustrate the same process of industrial mutation – if I may use that biological term – that incessantly revolutionizes the economic structure *from within,* incessantly destroying the old one, incessantly creating a new one. This process of Creative Destruction is the essential fact about capitalism. It is what capitalism consists in and what every capitalist concern has got to live in.

According to Schumpeter, creative destruction is not primarily an adaptive reaction to something happening within the production systems. Rather, the forces of innovation mainly are exogenous and come from outside the system. Schumpeter also takes the position that innovations by entrepreneurs are creative but also destructive to an economic system that initially can be in equilibrium. Kirzner (1999) advocates the view that the entrepreneur and his ability to create and implement innovations repair a market system in disequilibrium so that it approaches an equilibrium situation.

It seems like Schumpeter changed his views on both innovation and creative destruction from his books about the economic development from 1912 and 1934 to publication of the book *Capitalism, Socialism and Democracy* in 1942. In early writing, the individual abilities of the entrepreneur to carry out the innovation process, from invention to innovation, and create a surplus in the market, are looked at as essential. Here, he was using the picture of the smaller individual enterprise and the underlying entrepreneur. In later writings (Schumpeter, 1942/1950), Schumpeter changed his view in the sense that he, to a much greater extent, viewed how larger companies could develop knowledge and gain profit from innovations. This also has consequences for how Schumpeter thinks about creative destruction; and in the literature about Schumpeter, one often uses the term Schumpeter Mark I for the early period when the ideal is small businesses with the individual innovative entrepreneur. Here, companies are small and to a very low degree are able to adapt to changing conditions and competition from new creative situations; so we get a birth and death process of small businesses, products, technologies and organizational forms.

The so-called Schumpeter Mark II logic is that larger companies have greater resources for research and development activities so that the companies to a much higher degree are able to adapt to new shocks in the market from creative intruders differently than small businesses. There have been several attempts to operationalize and test whether it is Schumpeter Mark I or Mark II which is best suited to explain historical data; see Andersen et al. (2006) and Wersching (2006). The more general conclusion is that Schumpeter saw a trend towards larger industrial units with more resources for innovation and adaptation to changing environments and also public sector interventions, and built this into his analysis.

RENEWED INTEREST FOR INNOVATION EMERGING IN THE LATE 1980s

Systems of Innovations

Different writers started to use the concept of 'systems of innovations' (SI) around 1990, including Lundvall (1992) and Nelson (1993). It started with the definition and use of this concept at the national level, 'national systems of innovations', and during the 1990s we got the definitions of and discussions about 'sectoral systems of innovations' (Breschi and Malerba, 1997) and 'regional innovation systems' (Cooke et al., 1997).

One important reason for introducing systems with innovation was to underline that innovation as a process normally did not take place in isolation. The SI approach asks us to specify the determinants of innovation processes and the relations between the components and institutions (like sets of common habits, norms, routines, laws, etc.) of the system. The case study later in this chapter fits into this method of developing the analysis where a steel construction-producing company is the organization and important components of the system are the different groups of the workforce, technologies and routines, relations to subcontractors, relations to buyers, etc.

In Fagerberg et al. (2005, p. 184), one strength of the SI approach is looked upon as:

> The SI approach places innovation and learning processes at the center of focus. This emphasis on learning acknowledges that innovation is a matter of producing new knowledge or combining existing (and sometimes new) elements of knowledge in new ways. This focus distinguishes the SI approach from other approaches that regard technological change and other innovations as exogenous.

This focus on system interdependencies and knowledge developments is a key element in the empirical study later in this chapter and one can say that the introduction of the SI approach paved the way for a closer link for innovation studies' relevance for the knowledge economy. On the other hand, the SI approach did not solve all problems for the analysis of innovations. The question about what is to be included in the system definition and what is to be left out has no simple solution. The reason for that seems to lie in the nature of the innovation concept. Innovations as a social process are dependent on the context and institutional set up. On the basis of the SI approach so far, we have not seen one final list of determinants that explains innovations in a coherent manner and that was not to be expected either. The SI approach has suggestions similar to other research reviewed in this chapter. The main contribution of the SI approach seems to lie in its focus on systematic dependencies and the importance on knowledge use and development.

Explanations and Measurement of Innovative Behavior

The systems of innovation approach contributes to a clearer understanding and definition of the innovation concept and also influences the discussion about how to measure innovations. Yet we cannot say that there has been a relatively consistent theory of innovation formulated and satisfying methods to measure it. Professor Karen R. Polenske at MIT states the following (Polenske, 2007, p. 5):

> I am struck by the many different interpretations of innovation, the lack of consensus on a framework both to define the theory of innovation and the way to measure it, and by the vast number of empirical studies that are done, but using relatively simplistic measures.

A starting point for an operational definition of innovation was given in what is called the 'Oslo Manual', the first version of the report published by OECD in 1992, since then revised several times, now released jointly by OECD and the European Commission (2005). The definition of innovation in the Oslo Manual is the following (OECD 1992):

> A technological product innovation is the implementation/commercialization of a product with improved performance characteristics such as to deliver objectively

> new or improved services to the consumer. A technological process innovation is the implementation/adoption of new or significantly improved production or delivery methods. It may involve changes in equipment, human resources, working methods or a combination of these.

The Oslo Manual deals with technological product and process innovations (TPP). The definition in the Oslo Manual implies a narrowing compared to Schumpeter's treatment and his definition of the term innovation. In addition to product and process innovations, Schumpeter also had market innovations, innovations because of the supply of new raw materials or intermediate products and innovations because of the new organizational structures. Another difference is that Schumpeter had clear ideas about what qualities were needed to define the innovator.

Throughout the 1990s, focus to a large extent was to find factors that actually contributed to product and process innovations. In the past ten years, one can see that there have been several studies with a widening of focus, more in the direction of how Schumpeter first defined innovation. In an article from 2006, Becheikh et al. (2006) present a systematic analysis of a large number of innovation studies from 1993 to 2003. In this article we see that many of the studies based their definition of innovation on the Oslo Manual or similar definitions. It is important to note that both the Oslo Manual and many of the studies referenced by Becheikh et al. (2006) define innovation as the dependent variable and the analysis is focused on how to find independent variables that can explain why we get (or fail to get) innovation activity.

The Oslo Manual also borrows from Schumpeter. First, it is generally agreed that the main characteristics of studying innovation have to do with processes. The start of the innovation process is often based on an invention or more generally – seeing new combinations, and then bringing it to market so it contributes to value creation. The normal way of looking at this is that the social surplus has to be connected to commercial profit that works through the price system. The majority of studies referenced by Becheikh et al. stress that innovation in a commercial sense must contribute to profit.

One problem with the definition of innovation in the Oslo Manual is the considerable degree of evaluation. For example, it talks about whether different production methods will be new or significantly improved compared to the previous production method used by the company. In Chapter 14 in this book, companies are asked about what a new production process means within the broader industrial mix: new to an individual firm, new compared to competing firms in the same country, new to the global production process, new in the sense that it requires new knowledge within the workforce, and so on. This question is also discussed by Becheikh et al. (2006) and they find many variants of answers.

One must also look at whether the innovation is considered to be radical or incremental. Many companies improve their production step by step and when this has been going on for a while, they say that the production method is significantly altered. But how measurements and assessments shall be carried out here, before the firm can conclude that the production process is significantly improved, is a matter of evaluation.

The study by Becheikh et al. (2006) gives a systematic account of the factors that affect the product and process innovation. In the article, we find a breakdown in what is called internal factors and external factors to explain innovation in companies. It results in seven internal factors and six contextual factors that are external to the company and several of these will be used in the empirical example later in this chapter. One can say that these categories represent a good overview of where to look for explanations of why innovation takes place. Some of the factors like company size, corporate culture and the company's strategy for knowledge acquisition are key variables in Schumpeter's discussion of how innovation can be explained.

We will briefly mention a study (Drejer, 2004), where the types of innovations are developed one step further than studies by Becheikh et al. (2006). Drejer (2004) looks at:

- expertise-field innovations;
- external relational innovations;

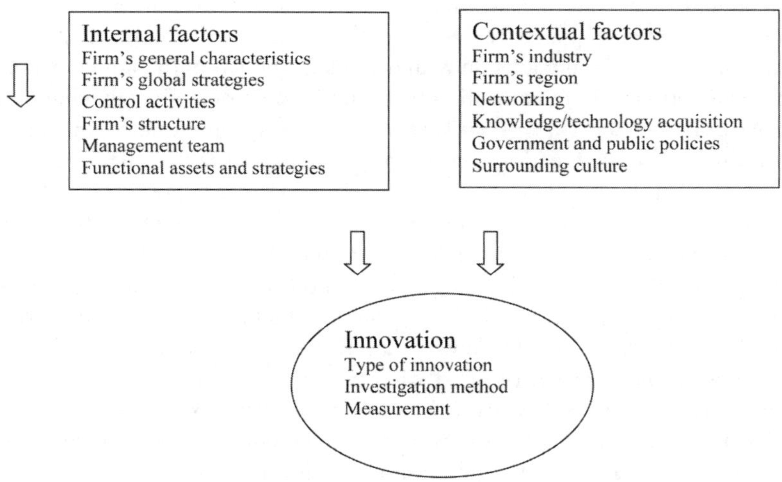

Figure 4.1 A framework for integrating innovation findings (Becheikh et al., 2006)

- innovations based on codifying and making explicit knowledge and processes which have hereto been informal and tacit.

What is very interesting here is the third category, which specifically addresses the processes of codifying knowledge as a source of innovation. This becomes particularly important in relation to the empirical example in this chapter.

An attempt to bring new developments in the analysis of innovation processes is provided by Chesbrough (2006) with the concept of open innovations. Chesbrough discusses the concept of open innovation and how this can contribute to previous theories and research on innovation processes. The open innovation concept emphasizes the importance of the research phase, both to take care of signals that come from within the system and to combine these with external research and knowledge input. Furthermore, one must accept in the development phase of an innovation that knowledge is 'leaking' out of the organization. In an open innovation world, companies that 'plan' innovations for one market can see spin-offs occur within other markets.

KNOWLEDGE AND INNOVATION

When we analyse innovation and knowledge, it is important to be aware of the content of the concepts we use. Innovation as a phenomenon is a process where the time aspect is important and thus there is reason to focus on how the process takes place. The same can be said about the concept of knowledge and often we split into stock and flow of knowledge. Many new innovation studies also look back to Schumpeter's earlier works (e.g. Nørager, 2009), where his interests are particularly focused on the skills of the entrepreneur. Schumpeter points out different types of skills that are needed to have a successful innovation process:

- inventor abilities;
- innovator/entrepreneur – who develops the product for the market;
- the manager/the organization designer – who is able to run the business making profit;
- the capitalists with financial resources and skills.

To achieve a professional and successful innovation process, as Schumpeter pointed out around 1912, one must have professional skills in all the roles and functions mentioned above.

There are many aspects where there are research results related to innovation and innovation processes. In the empirical application found later in this chapter, we will specifically look at the relationships between knowledge and innovation. Nelson and Winter (1982) provide a thorough discussion of the distinction between tacit and explicit knowledge, building on Polanyi and how the distinction between explicit and tacit knowledge can be used in the analysis of the company's innovative activity. They also pick up from Schumpeter the emphasis on how procedures can be developed in the context of innovations; where process innovations are used as an example. The important conclusion is that tacit knowledge among employees stimulates innovations taking the form of procedures to be changed or done in entirely new ways. Nelson and Winter point out that an instructor (like a production group leader) can use his explicit knowledge to explain how a routine should be performed, but it is often individuals with particular skills that come up with innovative ideas. Such ideas are often based on the tacit knowledge of the individual.

From Polanyi's point of view tacit knowledge is not looked at as a reservoir but as a process. Polanyi (1958, 1966) more often speaks about tacit knowing than tacit knowledge. Polanyi develops the idea that tacit knowledge is a kind of 'background layer' in the mind of those who participate in production. These 'layers' are built up by skilled work, social interactions and language developments, formal education and other factors (see Jones and Miller, 2007). Under certain circumstances these 'layers' can appear as explicit knowledge in the sense that the workers find new or improved combinations that change routines within the production processes in ways that create value for the company. In the empirical part of this chapter we will refer to a study showing an example of this.

EMPIRICAL EXAMPLE

Project Background

Aker Verdal is a Norwegian company that produces equipment for the offshore oil sector. The North Sea has been their primary geographic market, but in recent years, Aker Verdal has produced equipment used in off-shore applications in Canada and the Gulf of Mexico. Between 2000 and 2009, the company had a total annual production value of about US$200 million and employed roughly 750 workers. The main output produced by Aker Verdal is steel jackets; a market which has experienced significant changes in demand, large fluctuations in volume and changes in design and engineering. It is a great challenge for the company to maintain continuity in production when activity is low and market fluctuations are the rule, not the exception.

Aker Verdal built a specific steel jacket called a Valhall between May 2008 and June 2009. The steel jacket weighed about 7000 tonnes, had a cost of roughly US$100 million, and took about 500 000 work-hours to build (see

Figure 4.2). The report from the Valhall project (Westeren, 2010) discusses many aspects of the production process; here, we will focus on the analysis and data collection focusing on the following questions:

- Were there process innovations during the production process?
- If so, what was the role of knowledge creation, transfer and use?
- What were the important internal and external factors affecting the innovation process?

Aker Verdal's core competency lies in the construction of steel jackets. The company offers total supply of this product. This means that the company has the necessary expertise to perform engineering with the necessary documentation, manufacturing, testing, final approval and marine installation.

The company has focused on welding technology and developing mechanized welding equipment and robots. One of the reasons for the company's success in the off-shore industry has to do with fulfilling very stringent requirements in welding quality for off-shore products in the oil sector. All welding equipment can be bought on the world market and the company has no technology-related patents. The competitive advantage for a company like

Figure 4.2 Valhall steel jacket in May 2009, just before final on-shore testing

Aker Verdal lies in the use of new cost-saving solutions. This requires an interaction in the entire production line, from design/drawings to manufacturing and work management and control/testing of the job. A technological advantage is therefore both the actual technological solution and in how the whole production system is organized.

Data Collection

Data collection for this project was based on observation and registration of central steps of the manufacturing process. The data collection followed closely the entire production process and the central concept in the manufacturing design of a steel jacket is referred to as the work package. The production of Valhall was divided into approximately 600 work packages and each work package had complete drawings of the 'piece' that was produced and a description of all work processes (including welding technology) and test procedures for the acceptance of the work package in its final form and an estimate of the number of work-hours required for the production process. Work packages differed in size and content; from small jobs with less than 100 work-hours to large scale operations of more than 2000 work-hours (see Figure 4.3).

The central characteristic of the data collection in the project was how the knowledge and innovation variables were linked to other properties of each work package. This was done by using an indicator set. For every indicator we registered values which in practice were done by having a research team member (at least) one day each week at the production site. Special attention was paid to situations where work-hours were saved or lost compared to the standard calculation for the specific work package. The most important function of how communication and knowledge transfer took place in the production of the work packages included the central position of group leaders.

The variables used in the project were first divided into three main groups:

1. background and structural variables;
2. variables related to knowledge capital in production;
3. variables related to results from production activities.

Background and structural variables:

- calculated number of work-hours for each work package;
- actual number of work-hours for each work package;
- size (actual number of work-hours) of the work package – recoded;
- changed the number of calculated hours of work package during production process;
- group size (number of welders and other workers at work package);

Figure 4.3 Details of the Valhall steel jacket, focusing on one work package called a mudmat. This work package took 2100 work-hours to build

- average age of workers at work package;
- average employment years at the company;
- number of non-Scandinavian-speaking workers.

Human knowledge capital variables:

- level of competence;
- participated in on-the job training;
- welding technology.

Structural knowledge capital variables:

- automated welding;
- understanding of drawings – level of complexity;
- measurement of accuracy – level of complexity.

Relational knowledge capital variables:

- participation with other firms.

Variables linked to outcome of production:

- extra hours used on measuring accuracy;
- extra hours used on interpreting drawings;
- production errors – extra hours spent;
- cooperating with other firms – extra hours spent;
- possibilities for process technology improvements – hours saved;
- loss of hours because of non-responsive behavior;
- hours lost because of problems with logistics;
- external quality control – extra hours;
- internal quality control – extra hours;
- waiting because of damaged equipment;
- not sufficient manpower;
- extra hours because of not sufficient manpower.

Analysis of Innovations During Production of the Valhall Steel Jacket

We have defined two variables that reflect process innovation activities related to manufacturing of work packages in the data collection.

Innovation variables:

- had an idea about process innovations;
- process innovations – work-hours saved.

It is difficult to differentiate between what can be looked upon as an improvement in the production process and what is to be labeled as an innovation. Production is organized in such a way that the group leaders receive drawings and all engineering information about the production of each work package from the engineering department. For steel constructions in the oil sector there are very strict requirements for how welding, assembling, measuring and other production activities shall be performed. Further, there is careful control, both internally (from the company) and externally (from the buyer). During the production process, drawings, instructions and production methods are closely followed.

But in some cases changes have been initiated in connection with the planning and production of the work package. Examples of process innovations we found included:

- where work operations were conducted in a different way than planned;
- when new welding techniques were used; or
- when the steel constructions were rotated (during production) in a different way than described in the planning material.

To register a process innovation it must meet the following requirements:

- the production process must be carried out in a different way than described in the planning material for the work package;
- this must also lead to a productivity increase – meaning that we could register the number of hours saved compared to what was originally calculated.

The first innovation variable was 'had an idea about process innovations'. This variable was defined in the data collection by picking out work packages where we registered an innovation according to the criteria mentioned above. The data collection included 29 work packages of 431 where this has happened. This occurred in about 7 percent of all work packages.

Several regression model specifications were done based on the variable 'had an idea about process innovations' as the dependent variable. Results suggested that the variable 'competence level' was stable and significant (at the 5 percent level) regardless of the model specification. The same considerations hold true for variables that reflect size of the work package. Whether 'actual number of work-hours for each work package' or a 'recoded size variable' (where the work packages are grouped) were used, size was significant in the regression models. Regression results for process innovation are presented in Table 4.1.

The variable 'extra hours used on interpreting drawings' was significant in the regression model, which could be interpreted that the work packages shown with special attention were those where we frequently found innovation activity. The variable 'extra hours used on interpreting drawings' was actually a problem variable from the company's point of view, meaning that either the explanations from the engineering department were not good enough or the group leader's explanation/communication was not sufficient. In any case, the extra attention the work package received fueled unexpected innovative actions that saved work-hours.

The variable 'not sufficient manpower' was also significant in the regression model with an expected negative sign. This result suggested that the less stress there was related to the production of work package, the better the chance that innovation activities took place. Also significant in the model was the variable 'share of non-Scandinavian-speaking workers' with robust t-values and negative sign on the coefficient. This implied the larger the share

Table 4.1 Regression results for the variable: had an idea about process innovations

	Non-standardized coefficients		Standardized coefficients		
	B	Standard error	Beta	*t*	Sig.
(Constant)	0.062	0.118		0.526	0.599
Level of competence	0.092	0.019	0.234	4.793	0.000
Size of the work package – recoded	0.015	0.006	0.140	2.665	0.008
Extra hours used on interpreting drawings	0.019	0.007	0.129	2.755	0.006
Not sufficient manpower	–0.120	0.055	–0.107	–2.194	0.029
Share of non-Scandinavian speaking workers	–0.004	0.001	–0.241	–4.618	0.000

R	*R* square	Adjusted *R* square	Standard error of the estimate
0.352[a]	0.124	0.113	0.237

Dependent variable: had an idea about process innovations

of foreign-speaking employees, the smaller the chance that innovative activities would occur. This conclusion suggested that communication was a central element to innovation.

In the regression model we found several interesting significant explanatory variables, but the explanatory power was limited by an adjusted *R* square of only 11.3 percent. The most obvious reasons why this occurred were that innovation is a term that was difficult to operationalize and there were likely to be other factors coming into play that were likewise difficult to operationalize and were not included in the analysis.

The variable 'process innovations – work-hours saved' included information about how many hours the company actually had saved because of the innovation that took place on the actual work package (see Table 4.2). Saved work-hours for all registrations came to 846 hours and the average savings was about 30 hours for each work package.

Table 4.2 Regression results for the variable: process innovations – hours saved

	Non-standardized coefficients		Standardized coefficients		
	B	Standard error	Beta	*t*	Sig.
(Constant)	6.609	5.208	–	1.269	0.206
Level of competence	1.618	0.775	0.129	2.089	0.038
Size of the work package – recoded	0.785	0.221	0.240	3.549	0.000
Extra hours used on interpreting drawings	0.444	0.254	0.100	1.751	0.081
Not sufficient manpower	–3.229	1.949	–0.101	–1.656	0.099
Share of non-Scandinavian speaking workers	–0.135	0.038	–0.256	–3.556	0.000
Average years of employment	–0.269	0.185	–0.094	–1.459	0.146

R	*R* square	Adjusted *R* square	Standard error of the estimate
.342	0.117	0.098	7.814

Dependent variable: process innovations – work-hours saved

The regression model with 'process innovations – work-hours saved' as dependent variable also had size of the work package as a significant explanatory variable in the sense that the larger the work package, the larger the chance to get a high number of hours saved because of process innovation. Also, the variable 'level of competence' had a positive coefficient, meaning that the higher the level of competence, the more hours saved. This result suggested a connection between knowledge and competitiveness.

The third variable that held at the 5 percent level in the regression model was 'share of non-Scandinavian-speaking workers' which entered with a negative sign and robust *t*-value. This can be interpreted to suggest that the lower the proportion of foreign-speaking workers, the greater the number of saved hours because of innovation.

Two variables from the previous analysis which were also included in the analysis of the number of hours saved by process innovation were 'extra hours used on interpreting drawings' and 'not sufficient manpower'. Both of these variables had the same sign as in the previous regression model, but now they were less significant (significant at the 10 percent compared to 5 percent in the previous model). This may be explained by the different construction of innovation variable. The second model took for granted that we had an innovation and analysed contributions to the larger saving of work-hours resulting from the innovation. Apparently, it was important to avoid stress in creating a favorable situation for increasing productivity from innovation. Furthermore, in the second model we used the variable 'average years of employment' in the company, but it was not significant.

CONCLUDING REMARKS

One link from Schumpeter's analysis of innovation and the knowledge economy has to do with finding new combinations. This has been and will continue to be the challenge in innovative activities like manufacturing industries, service production, in the public sector and elsewhere. One can discuss Schumpeter's view that the most important source for the ideas behind the new combinations is coming from outside the system; we also see contributions from inside of firm and research systems now. But still we have the problem of building a more coherent theory of innovations so we can better explain innovative activities. It is interesting to note that the efforts to expand the more fundamental platform of innovation theory still is progressing, one example being *The Microtheory of Innovative Entrepreneurship* by William J. Baumol (2010).

In the empirical application, we analysed and found relationships between knowledge and innovation. A consequence of this should be to go further into the concept of skills and their relation to innovation activities. This underlines the role tacit knowledge plays in the context of innovation. An interesting research subject in the project work was the analysis of tacit knowledge. The empirical study here showed that it was not possible to have an overview of how the process of converting tacit to explicit knowledge took place in advance, but it is possible for a firm to have ideas about what factors contribute to an innovative milieu, thus converting tacit knowledge into innovative solutions.

This supports Polanyi's view that tacit knowledge should not be looked at as a reservoir but as a process. Polanyi more often speaks about tacit knowing than tacit knowledge. Polanyi develops the idea that tacit knowledge is a kind of 'background layer' in the mind of those who participate in production. These 'layers' are built up by skilled work, social interactions and language developments, formal education and other factors. Under certain circumstances, these 'layers' can appear as explicit knowledge in the sense that workers find new or improved combinations that change routines within the production processes in ways that create value for the company. This is substantiated with empirical evidence. Further, results suggest that the interplay between knowledge creation, the use of different knowledge channels and communication variables play an important role in creating innovative solutions. The primary contribution of this chapter dealt with bringing together different elements from theory with thorough data collection and analysis of firm behavior, thus yielding new insights for understanding the phenomena of innovation.

REFERENCES

Andersen, E.S., M.S. Dahl, B.-Å. Lundvall and T. Reichstein (2006), Schumpeter's process of creative destruction and the Scandinavian systems: A tale of two effects, Paper for the DRUID Conference, 18–20 June 2006.

Baumol, William J. (2010), *The Microtheory of Innovative Entrepreneurship*, Princeton: Princeton University Press.

Becheikh, N., R. Landry and N. Amara (2006), 'Lessons from innovation empirical studies in the manufacturing sector: a systematic review of the literature from 1993–2003, *Technovation*, **26** (4–5), 644–64.

Breschi, S. and F. Malerba (1997), 'Sectoral innovation systems: technological regimes, Schumpeterian dynamics and spatial boundaries' in C. Edquist (ed.), *Systems of Innovation: Technologies, Institutions and Organizations*, London: Pinter.

Chesbrough, Henry (ed.) (2006), *Open Innovation*, Oxford: Oxford University Press.

Cooke, P., U.M. Gomez and G. Etxebarria (1997), 'Regional systems of innovation: institutional and organisational dimensions', *Research Policy*, **26**, 475–91.

Drejer, I. (2004), 'Identifying innovation in surveys of services: a Schumpeterian perspective', *Research Policy*, **33**, 551–62.

Fagerberg, J., D.C. Mowery and R.R. Nelson (2005), *The Oxford Handbook of Innovation*, Oxford: Oxford University Press.

Jones, B. and B. Miller (2007), *Innovation Diffusion in the New Economy*, London: Routledge.

Kirzner, Israel M. (1999), 'Creativity and/or alertness: a reconsideration of the Schumpeterian entrepreneur', *Review of Austrian Economics*, **11**, 5–17.

Lundvall, B.-Å. (ed.) (1992), *National Systems of Innovation: Towards a Theory of Innovation and Interactive Learning*, London: Pinter.

Nelson, R.R. (ed.) (1993), *National Systems of Innovation: A Comparative Study*, Oxford: Oxford University Press.

Nelson, R.R. and S.G. Winter (1982), *An Evolutionary Theory of Economic Change*, Cambridge, MA and London: Harvard University Press.

Nørager, Michael (2009), 'How to manage SMEs through the transformation from non-innovative to innovative?', Copenhagen Business School, Department of Management, Politics and Philosophy, PhD Series 19.

OECD (Organisation for Economic Co-operation and Development) (1992), *Proposed Guidelines For Collecting And Interpreting Technological Innovation Data*, Oslo Manual, Paris: OECD.
OECD (Organisation for Economic Co-operation and Development) (2005), *Proposed Guidelines for Collecting and Interpreting Technological Innovation Data*, Oslo Manual, third (revised) edition, Paris: OECD.
Polanyi, M. (1958), *Personal Knowledge*, Chicago: University of Chicago Press.
Polanyi, M. (1966), *The Tacit Dimension*, New York: Doubleday.
Polenske, Karen R. (ed.) (2007), *The Economic Geography of Innovation*, Cambridge: Cambridge University Press.
Schumpeter, J.A. (1926), *Theorie der wirtschaftlichen Entwicklung* (2nd edn), Munich and Leipzig: Duncker and Humblot.
Schumpeter, J.A. (1934), *The Theory of Economic Development: An Inquiry into Profits, Capital, Credit, Interest and the Business Cycle*, London: Oxford University Press.
Schumpeter, J.A. (1942/1950), *Capitalism, Socialism and Democracy*, 3rd edn, New York: HarperCollins.
Wersching, K. (2006), *Schumpeterian Competition, Technological Regimes and Learning through Knowledge Spillover*, Department of Business Administration and Economics, Bielefeld University.
Westeren, K.I. (2010), *Kunnskap og konkurranseevne hos Aker Verdal*, Utredning nr 119, Nord-Trøndelag University College, Norway.

5. Star scientists and regional knowledge transfer

Michaela Trippl and Gunther Maier

INTRODUCTION

In recent years, there has been a growing recognition that highly skilled researchers and scientists are a key driving force for regional knowledge-based development, growth and technological change (see, for instance, Horowitz, 1966; Thorn and Holm-Nielsen, 2008, Trippl and Maier, 2011a). The knowledge generated and accumulated by these individuals and the expertise they possess are seen to constitute essential inputs for industrial innovation processes, particularly for those in science-based industries (Pavitt, 1984). In the meantime the increasing significance of knowledge linkages between universities and the economy and the involvement of scientists in commercialization processes of scientific findings is well documented and vigorously discussed in the academic literature (Mowery and Sampat, 2005; Gunasekara, 2006; Bergman, 2010). However, so far only a few scholars have focused on world-class researchers and investigated the nature and extent of their knowledge sharing activities (Zucker et al., 1998a, 1998b, 2002; Furukawa and Goto, 2006; Baba et al., 2009; Schiller and Revilla Diez, 2010). While their studies have provided interesting insights into the role of elite scientists in promoting regional innovation and high-tech development, little is still known about the relative importance of different types of intraregional knowledge transfer mechanisms which are crucial in this regard. Moreover, there is a limited understanding of those factors which have an influence on whether elite scientists employ various kinds of knowledge transmission.

The key aim of this chapter is to enhance our understanding about these core issues by exploring intraregional knowledge transfer activities performed by star scientists. Star scientists are defined here as the world's top and most renowned researchers, identified by the number of citations they generated in journals in the ISI database. Based on the findings of a worldwide web-based survey of 720 star scientists in 21 different scientific disciplines, we examine whether and in what ways star scientists are engaged in regional knowledge transfer and which factors influence the degree and intensity of such activities.

More specifically, this chapter deals with the following main research questions. To what extent do star scientists engage in regional knowledge sharing activities and what is the relative importance of different modes of knowledge transmission employed by stars in this context? What are key factors determining whether or not star scientists engage in intraregional knowledge transfer activities and which factors influence the intensity of different modes of knowledge exchange?

The remainder of this chapter is organized into four sections. The second section provides a short overview on the role that top scientists and elite researchers play in regional development and it identifies a set of mechanisms by which star scientists might share their advanced knowledge and expertise with other regional actors. Section three contains the empirical part of this chapter. After discussing the methodology, the data of our research and some key sample characteristics, we report the findings of the descriptive analysis of star scientists' knowledge transfer activities in their regions of choice. Then we present statistical models measuring the influence of a set of factors on forms and intensity of intraregional knowledge-sharing activities performed by the surveyed star scientists. Finally, in the fourth section we summarize the key findings of our empirical analysis and draw some conclusions.

CONCEPTS AND PREVIOUS LITERATURE

In the knowledge-driven economy, top academics and outstanding researchers are acknowledged to play a critical role in spurring regional high-tech development, science-based innovation and growth (Horowitz, 1966; Zucker and Darby, 2006; Thorn and Holm-Nielsen, 2008). In recent years, a considerable body of work has dealt with the rise in importance of university–industry partnerships, increasing attempts to convert high-quality scientific findings into commercial success, and the involvement of academics in industrial innovation processes (see, for example, Mowery and Sampat, 2005; Gunasekara, 2006; Bergman, 2010). Enhancing knowledge linkages between scientists and economic actors has also become a core goal of innovation and technology policies in many countries and regions (see, for example, Etzkowitz and Leydesdorff, 2000; Feldman and Owens, 2007). Indeed, around the world universities and scientists face growing expectations to contribute to industrial innovation and place their knowledge at the commercial disposal of companies. Recent empirical studies (Agrawal and Henderson, 2002; Van Looy et al., 2004; Breschi et al., 2007; Calderini et al., 2007; Lowe and Gonzales-Brambila, 2007; Stephan et al., 2007; Azoulay et al., 2009) have shown that there is no trade-off between academic productivity of top scientists and their participation in commercialization activities of research findings. On the contrary, a complementary relationship between scientists' high quality academic research and their involvement in industrial innovation, patenting and new firm formation seems to exist.

In this chapter we focus on star scientists, that is, the world's most renowned academics and research professionals. We identify stars by the number of citations they generated in journals in the ISI database (see also p. 80). Star scientists possess unique cutting-edge knowledge and they make excellent contributions to the advancement of science and technology in their respective disciplines. However, empirical evidence about intraregional knowledge-sharing processes involving these elite researchers is scarce. Only a few studies explored the role of star scientists in regional development and their knowledge transfer activities. Zucker and colleagues showed in the rapidly advancing area of biotechnology that star scientists making major discoveries play an important role, influencing the use of the new technology by firms (Zucker and Darby, 1996, 2001, 2006; Zucker et al., 1998a, 1998b, 2002). Zucker and Darby (1996) and Zucker et al. (1998b) introduced the concept of biotechnology stars based upon productivity measured by the number of articles written through the 1990s which reported a genetic-sequence discovery. Direct involvement of these stars proved to be a major factor in determining which firms were ultimately major winners in biotechnology (Zucker et al., 1998b, 2002; Zucker and Darby, 2001). In another paper Zucker and Darby (2006) extend the concept of star scientists to all areas of science and technology. They demonstrate that the number of stars in a US region or in one of the top 25 science and technology countries has a consistently significant and quantitatively large positive effect on the probability of firm entry in the same area of science and technology. These findings lead them to conclude that the stars themselves, rather than their potentially disembodied discoveries, play a crucial role in the formation or transformation of high-tech industries, emphasizing their embodied knowledge, insight, taste and energy. The physical presence of star scientists thus matters, as it has an impact on the formation and transformation of high-tech firms. The evidence provided by Zucker and Darby (2006) strengthens the case for the importance of the work of these extraordinary individuals for the economic development of regions and nations. In a similar vein, a recent study by Schiller and Revilla Diez (2010) highlighted that star scientists in Germany are rather strongly involved in regional knowledge transfer. Indeed, many of the activities performed by Germany's top scientists – particularly scientific collaborations, new firm foundation, recruiting staff and PhD students – were found to be strongly localized in nature. Collaborations between star scientists and companies, in contrast, proved to be weaker at the regional level. Looking at European-based stars Trippl and Maier (2011b) have found that these outstanding scientists employ many different mechanisms to transfer their knowledge to regional actors. Finally, Trippl (2011a) has shown that star scientists do not only create intraregional but also international knowledge linkages.

The empirical findings on star scientists recapitulated above have contributed to a better understanding of the role of world-class researchers in regional knowledge circulation and innovation. However, more empirical evidence and additional research are required to get deeper insights into the extent of intraregional knowledge-sharing activities performed by star scientists and the relative importance of different knowledge transfer mechanisms which matter in this context. Moreover, little is known about which factors determine the extent of their engagement in intraregional knowledge transfer and influence the intensity of different modes of knowledge exchange.

In the following, we seek to be more specific about the mechanisms by which top researchers and star scientists potentially engage in intraregional knowledge transfer. We do not look at unintended spillovers which may result from the mere presence of star scientists in a region. Such spillovers do not require any form of engagement or activities by the top researchers and might thus be observable even for isolated star scientists who lack any connections to actors and organizations at regional level. We do not argue that such unintended spillovers cannot play an important role for regional development and innovation. Nevertheless, in this chapter we only focus on conscious knowledge transfer activities performed by star scientists in their region of choice. By creating knowledge linkages to regional actors, star scientists embed themselves in their respective regions. In our attempt to conceptualize the role of embedded star scientists we draw a distinction between the regional academic, industrial and policy world. Drawing on the work done by Keeble (2000), Tödtling et al. (2006) and Schiller and Revilla Diez (2010) we suggest the following typology of mechanisms or forms of engagement (Figure 5.1) by which star scientists might share their knowledge with other regional actors (see also Trippl, 2011a; Trippl and Maier, 2011b). The components of Figure 5.1 include the following.

- Regional academic world: star scientists can play a powerful role in knowledge diffusion within the regional science system. A key mechanism in this context is related to the educational function of star scientists and their contribution to the dynamic evolution of the regional scientific labor market. Top scientists are considered to play an essential role in attracting the best young talents and to guide them into fruitful research areas (Laudel, 2005). If these talents do not move away after having finished their studies, but continue to stay in the region to work for other research organizations, we might expect positive effects for the regional academic world. Another crucial channel for knowledge exchange is associated with scientific collaborations. The more cooperative linkages star scientists maintain with other researchers present in the region, the more vividly will their knowledge circulate at the regional level.

- Regional industrial world: star scientists may also establish various connections to the companies located in the region, thus supplying their knowledge to the corporate world. We distinguish between the following mechanisms or channels in this respect. First, star scientists might act as a source of highly qualified workers for regional companies. The mobility of highly trained graduates who have been educated and monitored by elite scientists from research institutions to firms represents a significant knowledge transfer channel, supporting the regional diffusion and commercial application of new scientific expertise derived from university research. Second, star scientists might also promote the entrepreneurial spirit and activities of their students in their region. Third, both formal and informal collaborations, particularly in the form of R&D projects, constitute another key channel for the exchange or transfer of knowledge between star scientists and regional firms. Fourth, selling patents to regional companies represents another core mode by which stars can transfer their knowledge to the industrial world. Fifth, top scientists

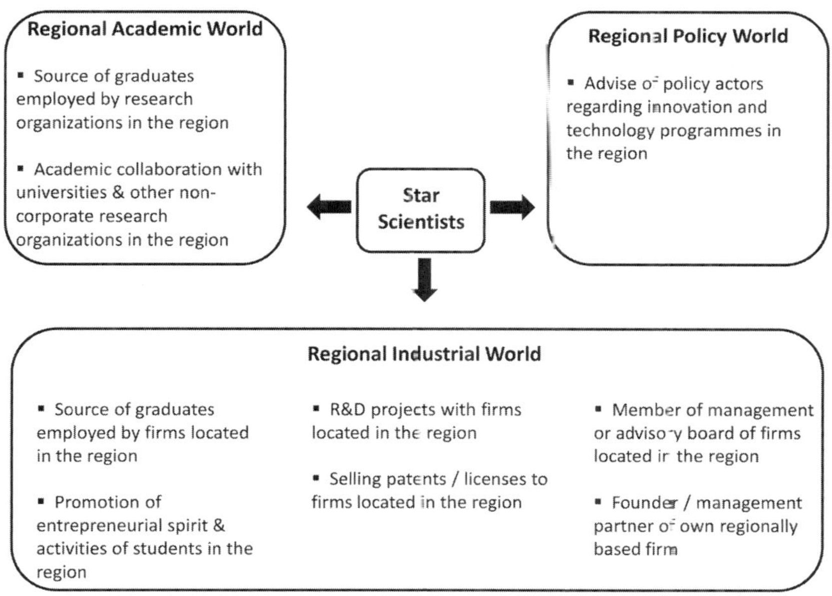

Source: Trippl and Maier, 2011b

Figure 5.1 Star scientists and intraregional knowledge transfer activities

might also engage in regional knowledge sharing by working part of their time as a member of the management or advisory board of regional companies. Sixth, we consider new firm foundation by stars as a specific mechanism for transmitting scientific knowledge to the industrial world. Arguably, the latter three modes of knowledge sharing constitute the most direct forms of commercializing top-level scientific expertise and require particularly high levels of engagement by star scientists.

- Regional policy world: top researchers might also transfer their knowledge and expertise to the regional policy world by advising public authorities, governments and policy agents with respect to the design of innovation and technology programs and measures.

In the next section we will present the findings of our empirical analysis. We will investigate the extent to which star scientists are engaged in regional knowledge sharing activities with other regional actors. We will analyse the relative importance of the different modes of knowledge exchange and examine which factors influence a star scientist's engagement in such activities.

EMPIRICAL ANALYSIS

Data and Methodology

The empirical results reported in this chapter stem from a worldwide web-based survey (conducted in August and September 2008) of so-called star scientists, that is, the world's top and most renowned scientists and research professionals. More precisely, making use of the database ISI Highly Cited, we define star scientists as authors of highly cited research papers. ISI Highly Cited is an online information service provided by the Institute for Scientific Information (ISI), a subsidiary of Thomson Incorporated. With this freely accessible website (http://isihighlycited.com/) one can identify individuals, departments and laboratories that made important contributions to the advancement of science and technology in recent decades. The importance of contributions is identified by the number of citations they generated in journals in the ISI databases. ISI Highly Cited distinguishes between 21 different areas of research (subject categories) such as clinical medicine, engineering, physics or social sciences and identifies approximately the 250 most cited individuals in each category. The information in ISI Highly Cited is based on publications and citations from the period 1981–2002. The quantitative approach and the extensive – but not very current – observation period represent a limitation in contrast to alternative (for example survey based, qualitative) approaches. Older researchers with an extensive publication record may have a better chance of being classified as star scientists because of the extensive observation period. Certain researchers and scientists who belong to the very top of their class are not included because they have not yet accumulated enough publications and citations (see also Trippl, 2011b).

Table 5.1 Sample characteristics I (percentage of star scientists)

		%
Gender (n = 720)	Female	5.6
	Male	92.6
	Missing	1.8
Year of birth (n = 720)	Mean	1946.5
Type of institution (n = 720)	University	70.4
	Non-university research entity	18.3
	Corporate research unit	2.1
	Other	5.8
	Missing	3.3
Extra-functional posts (n = 720)	University professor	45.6
	President of association/committee	17.9
	Entrepreneur	11.4
	University board member	11.8
	Director of non-university research entity	7.4
	Member of firm advisory board	9.0
	Member of firm board of directors	8.5
	University lecturer	3.5
	Other	16.0
	None	21.5
Type of research (n = 720)	Exclusively fundamental research	21.5
	Mostly fundamental research	29.0
	Rather fundamental research	11.3
	Fundamental and applied research	22.1
	Rather applied research	4.7
	Mostly applied research	8.1

		%
	Exclusively applied research	1.7
	Missing	1.7
Research discipline (n = 720)	Natural sciences	53.3
	Agriculture science	3.1
	Engineering and technology	10.1
	Medical and health sciences	23.3
	Social sciences	7.6
	Missing	2.5
Mobility background (n = 720)	Non-movers	47.9
	Expatriates	25.1
	Returnees	26.9
Expatriates: years already spent abroad (n = 181)	Mean (min. 0.7, max. 60): 29.5	
	1–10 years	11.6
	11–20 years	8.8
	21–30 years	29.8
	31–40 years	30.9
	More than 40 years	16.6
	Missing	2.2
Returnees: years spent abroad (n = 194)	Mean (min. 0.5, max. 40): 5.9	
	Less than 1 year	1.6
	1–3 years	49.0
	4–10 years	32.5
	More than 10 years	12.9
	Missing	4.1

The database includes approximately 5600 star scientists, comprising less than 0.5 percent of all worldwide publishing researchers. We contacted all 3274 star scientists who provide their contact information (e-mail address) in the database and invited them to fill in our questionnaire. Only 433 of them were not reachable due to invalid e-mail addresses. Out of the remaining 2841 star scientists, we have received 720 questionnaires, yielding a response rate of 25.34 percent.

Table 5.1 provides an overview on key characteristics of the overall sample. The overwhelming majority of responding star scientists are male (92.6 percent) and more than 50 percent are older than 60 years, signaling that they are at a relatively mature stage of their professional careers. A vast majority of the sampled scientists (70 percent) are employed by universities. About 18 percent are working for non-university research institutions, while the proportion of star scientists from the corporate sector is negligible (2 percent). Under the category 'other', the respondents have indicated that they are retired, have founded their own company, work for the government, or do non-profit research or consulting. A great majority of respondents (46 percent) hold a professorship in addition to their main occupation as an extra-functional post. Considering that 70 percent of the sampled star scientists were employed by universities, we assume that these 46 percent include additional professorships such as full professorships, while primarily working for an independent university research center (distinguished professor or endowed chair), adjunct professorships or visiting professorships at other (foreign) institutions as well as honorary professorships. Almost 22 percent of star scientists stated that they do not hold any extra-functional posts at all. The top researchers included in our sample are strongly oriented towards fundamental research. About 50 percent of star scientists stated that they exclusively (22 percent) or mostly (29 percent) carry out fundamental research. Another 22 percent do both. Looking at the subject areas, we found that biology and biochemistry (8.8 percent), chemistry (8.2 percent) and ecology and environment (7.4 percent) represent the most important research disciplines. Interestingly, none of the star scientists working in law and education has responded.

If we classify the 21 subject categories according to the *Frascati Manual* (OECD, 2006) into broader fields of science and technology, we can see that more than 50 percent of the respondents are working in the field of natural science, and another 23 percent in medical and health sciences. Other categories (engineering, social science, agricultural science) play a minor role in comparison. Finally, we also collected data on the mobility background of the surveyed star scientists. Almost 48 percent of them can be classified as non-movers. Non-movers are defined here as scientists who have, so far, not relocated internationally for professional purposes, but have stayed in their home countries. Another 52 percent have an international mobility background. We can draw a distinction between expatriates on the one hand and returnees on the other. Expatriates are defined here as researchers who have left their home countries and now live and work at a foreign location. Their share in the sample is 25 percent. On average they have already spent 30 years away from home. Returnees (scientists who have returned to their home countries after living abroad for a substantial period of time) represent 27 percent of all sampled stars. They have spent on average six years abroad, before relocating back home.

Table 5.2 Sample characteristics II (location pattern of star scientists)

	Number of stars	%
Top regions		
New York (USA)	35	5.4
Los Angeles/San Diego (USA)	32	5.0
San Francisco Bay area (USA)	29	4.5
Washington/Baltimore (USA)	29	4.5
London (UK)	26	4.0
Boston (USA)	21	3.3
Dallas/Houston/San Antonio (USA)	19	3.0
Chicago/Milwaukee (USA)	14	2.2
East Anglia (UK)	12	1.9
Raleigh Durham (RTP) (USA)	12	1.9
	229	35.6
Followers		
Tokyo (Japan)	11	1.7
Kyoto (Japan)	9	1.4
Ann Arbor (USA)	8	1.2
Boulder (USA)	8	1.2
Munich (Germany)	8	1.2
Philadelphia (USA)	8	1.2
Seattle (USA)	8	1.2
Atlanta (USA)	7	1.1
Copenhagen (Denmark)	7	1.1
Glasgow/Edinburgh (UK)	7	1.1
	81	12.6
Other regions		
(each hosting fewer than 7 stars)	334	51.9
Total	644	100.0
(Missing: 76)		

The surveyed star scientists are unevenly distributed in space. The USA clearly has the lead, hosting 56 percent of all elite researchers included in our sample. Europe accounts for 28 percent of all star scientists, while Asia

(7 percent), Canada (4 percent), Oceania (4 percent) and other parts of the world (1 percent) show a limited capacity to attract or retain top researchers in comparison. Looking at the location pattern of stars at the regional level, we can observe that the leading ten regions host not fewer than 35.6 percent of all top researchers included in our sample. As revealed in Table 5.2, there is a clear dominance of US regions with internationally renowned universities such as New York, Los Angeles/San Diego, San Francisco Bay area, Boston, and so on. Only two regions in Europe (London and East Anglia) are among the top ten ranked locations. Star scientists, thus, are strongly concentrated in only a few regions in the world (see also Trippl, 2011b).

Descriptive Analysis

A series of questions in our survey were intended to answer the main research question of this chapter: whether and in which ways star scientists engage in regional knowledge-sharing activities. Since our survey was intended for all star scientists worldwide, we could not provide a precise definition of the region. Thus, respondents were asked to indicate the particular sub-national region where they are currently located (US states, Germany's Bundesländer, etc.).

The first question we dealt with targets the star scientists' perception toward regional development oriented activities. We asked them to what extent they agree or disagree with the statement: 'Scientists and research professionals should play an active economic role in the regions where they are located'. Generally, the surveyed top scientists seem to have a rather positive attitude towards contributing to regional development. Almost 60 percent strongly or rather strongly agree with this statement.

But, does this positive attitude towards engagement in regional development activities become manifested in real actions? To what extent do star scientists engage in regional knowledge transfer activities at their current location and what is the relative importance of different mechanisms in this respect? Our results suggest that a large majority of the surveyed top researchers can be classified as embedded star scientists, as they exhibit various connections to other actors and organizations in their region (Table 5.3).

This is most obvious for linkages with other members of the scientific community in their respective region. Almost all star scientists (99 percent) collaborate with academic institutions at the regional level; more than 60 percent of them even do so in a quite strong way. A considerable share of star scientists (87 percent) also report that some of their former students are employed by research organizations in the region, pointing to their role in providing talented graduates for the regional scientific labor market. This holds true in particular for those 23 percent, who state that many or almost all of their former students have moved to other research organizations in the region.

Table 5.3 Types and intensity of regional sharing activities (percentage of star scientists)

	Total (n = 720)	Strong	Weak
Academic world			
Academic collaboration	98.6	64.4[a]	34.2[b]
Source of talent for scientific labor market	87.0	22.8[c]	64.2[d]
Industrial world			
Source of highly qualified workers	73.2	15.3[c]	57.9[d]
Fostering entrepreneurial spirit of students	73.2	17.8[a]	55.4[b]
R&D projects with firms	77.0	26.2[a]	50.8[b]
Selling patents to firms	33.9	4.6[a]	29.3[b]
Entrepreneur	14.3		
Member of firm board	26.6		
Policy world			
Advice of policy makers	72.4	14.8[a]	57.6[b]

[a] Strong: regular or frequent
[b] Weak: seldom or occasional
[c] Strong: a lot or almost all
[d] Weak: a few or some

However, the sampled top researchers do not only exchange knowledge with scientific actors in their regions of choice. We found evidence that they also adopt various mechanisms to share their advanced knowledge and expertise with regional firms. Key modes in this context include R&D partnerships with firms located in the same region (77 percent), the provision of highly skilled graduates (73 percent), and the promotion of the entrepreneurial spirit and activities of their students in their respective regions (73 percent). However, it should also be mentioned that more than 50 percent of all star scientists use these three modes in quite sporadic and weak ways. Some star scientists are also directly engaged in the commercialization of their scientific discoveries by selling patents (34 percent) to regional firms (however, only 5 percent carry out this activity regularly or frequently), acting as a member of the management or advisory board of regional firms (27 percent) or even running their own regionally based business (14 percent). Finally, our results also demonstrate that the surveyed top researchers are involved in shaping the institutional framework conditions at the regional level. Not fewer than 72 percent state that they advise regional policy makers regarding innovation and technology programs. Even 15 percent seem to be strongly involved in such territorial policy processes. Consequently, there are a large variety of mechanisms adopted by star scientists to share their knowledge with regional actors. It is not just the science system which seems to benefit from the physical presence of top researchers. Apparently, some researchers also maintain different kinds of linkages to regional firms or even have established their own firms, thus supplying their expertise to the industrial world. Finally, there is also evidence that their advanced knowledge and insights are incorporated in public programs geared towards promoting regional innovation and technology development.

Based on these results we can answer our first research question: Star scientists employ a large set of different mechanisms of regional knowledge transfer. Investigating their relative importance, we found that academic collaboration within the region is almost ubiquitous, closely followed in level by providing talent for the scientific labor market. That these academic core activities are so widely performed could have been expected. However, interactive activities in relation to regional firms and policy makers are also rather common. The more general activities of providing students for companies and fostering students' entrepreneurial spirit are performed by over 70 percent of star scientists. Similar shares also engage in more specific activities like performing R&D projects with firms and providing policy advice. But even rather specific activities like selling patents to firms and being a regional entrepreneur or a board member are reported by a substantial number of these highly qualified scientists. A look at the column 'strong' in Table 5.3 confirms the conclusion that star scientists are strongly embedded in their regional economies and by no means detached inhabitants of the academic ivory tower.

Statistical Models

Who are the scientists who engage in intraregional knowledge transfer and what types of star scientists are more likely to be strongly involved in such activities? Are these activities more likely in some disciplines than in others, do their past experiences or their stated preferences play a crucial role? These are issues which are at the core of our second research question that we will deal with in this section. In order to isolate the individual effects, we use multivariate models with indicators of the star scientists' engagement or level of engagement in the various knowledge transmission activities as dependent variables and their characteristics as explanatory variables. Since our dependent variables are either binary indicators representing whether or not the respective researcher engaged in this activity or discrete indicators of the intensity of their activities, we have to use discrete models. In the case of binary indicators we employ the binary logit model. In the case of indicators measuring intensity, we use the ordered logit model. In a number of cases we can aggregate intensity indicators into a binary variable and consequently apply both methods. The dependent and explanatory variables for our modeling exercise are taken from the variables that we discussed previously (see pp. 81ff and 85ff). The basic dependent variables are characterized in Table 5.3 and summarized in Table 5.4.

Table 5.4 Dependent variables

	Model I Binary (yes–no)	**Model II** Ordinal (intensity)	**Modell III** Binary (Strong)
Academic world			
Academic collaboration	AC_B	AC_O	AC_S
Source of talent for scientific labor market	SL_B	SL_O	SL_S
Industrial world			
Source of highly qualified workers	IL_B	IL_O	IL_S
Fostering entrepreneurial spirit of students	FS_B	FS_O	FS_S
R&D projects with firms	RD_B	RD_O	RD_S
Selling patents to firms	PA_B	PA_O	PA_S
Entrepreneur	EN_B		
Member of firm board	FB_B		
Policy world			
Advice of policy makers	PO_B	PO_O	PO_S

We can use most of them in three different ways:

1. as a binary variable, which is 1 when the respondent engages in the respective activity and 0 if not;
2. as an ordinal variable, which gives the intensity of the activity with higher values indicating higher intensity;
3. as a binary variable, which is 1 when the respondent's answer is in the top two intensity levels, 0 otherwise.

The correspondence between the variables, their respective versions and the variable names used is given in Table 5.4 above. These variable names will be used in the presentation of the results below. Table 5.5 presents the explanatory variables which were used for all models. We took into account a broad range of factors, including the international mobility background of the surveyed top

Table 5.5 Explanatory variables

Variable name	Variable description
Returnee	A dummy variable which is 1 when the respondent is a returnee, 0 otherwise
Expatriate	A dummy variable which is 1 when the respondent is an expatriate, 0 otherwise
Institution_nuru	A dummy variable which is 1 when the respondent is employed in a non-university research unit, 0 otherwise
Institution_corp	A dummy variable which is 1 when the respondent is employed in a corporation, 0 otherwise
Institution_other	A dummy variable which is 1 when the respondent is employed in another non-university unit, 0 otherwise
Type research	An indicator which is larger the more applied the respondent's research is
Science_Groups_Nat	A dummy variable which is 1 when the respondent is in natural science, 0 otherwise
Science_Groups_Eng	A dummy variable which is 1 when the respondent is in engineering, 0 otherwise
Science_Groups_Med	A dummy variable which is 1 when the respondent is in a medical discipline, 0 otherwise
Science_Groups_Agr	A dummy variable which is 1 when the respondent is in agricultural sciences, 0 otherwise
Economic_role	An index measuring whether or not and how strongly the respondent agrees that scientists should play an active economic role in their region (1 = strongly disagree, 5 = strongly agree)
Gender	A dummy variable which is 1 for a female respondent, 0 otherwise
Age	The respondent's age in years
Age_2	The respondent's age in years squared (divided by 1000) to allow for a U-shaped or inverse U-shaped relationship
Inst._context	A dummy variable which is 1 when the respondent's current location is in the USA, 0 otherwise
Socialization	A dummy variable which is 1 when the respondent's education was in the USA, 0 otherwise
Early_socialization	A dummy variable which is 1 when the respondent was born in the USA, 0 otherwise.

researchers, the type of institution they are working for, the degree of applied research, the scientific discipline, their general attitude towards an engagement of researchers in regional development, their gender and their age. Finally, we also included three variables, measuring whether being located or socialized in the USA makes a difference.

In the following, we discuss the key results of our statistical analyses. Generally, all three models employed here are rather weak in explaining the existence and intensity of linkages between the surveyed star scientists and actors from the regional academic and policy world (see Tables 5.6 to 5.8). The multivariate analyses provide more insights into those factors which influence the pattern of interaction between stars and the industrial world. The binary logit model (model I) presented in Table 5.6 shows that there are two key variables which determine whether or not the investigated top researchers adopt different modes of knowledge exchange with regional firms. We can observe that the type of research and the general attitude regarding the role of researchers in regional development have a strong and positive influence on the existence of all six types of knowledge sharing with the regional industrial world. The more applied the research done by stars, and the more positive their attitudes towards playing an economic role, the likelier it is that they transfer knowledge to the business community. Furthermore, for the activities 'R&D partnerships with firms' and 'selling patents' we also found that the scientific discipline matters. Star scientists working in natural sciences, engineering and medial and health sciences are more likely to engage in these two modes of knowledge transfer than stars working in the field of social science. Using an ordered logit model (model II, Table 5.7) we also studied the influence of the explanatory variables on the intensity of knowledge interactions between star scientists and the regional business community. Again, the type of research and the general attitude towards regional engagement seems to matter. Top researchers who are active in applied research and wish to take on an economic role are more intensively engaged in providing graduates to the labor market, fostering the entrepreneurial spirit of students, establishing R&D partnerships and selling patents than other stars. Moreover, we found a strong positive influence of specific science groups (natural sciences, engineering, medical and health sciences) on the intensity of R&D partnerships and selling patents. Finally, we examined which factors determined whether or not stars had strong links at the regional level, by employing another binary logit model (model III, Table 5.8). The results are similar to the findings reported above. A high degree of applied research and a positive attitude towards playing an economic role as researcher turned out to have a positive influence on the existence of strong relationships between stars and regional companies. This holds true for all four types of knowledge exchange included in this model. Furthermore, the same factors seem to be responsible for strong ties to the policy world. In sum, the three models presented above support some of our expectations concerning those factors which determine the embeddedness of top researchers into their respective regions. Three key variables had explanatory power; these being the general attitude about the role of researchers in economic development, a focus on applied research and the scientific discipline. To find a positive impact of these factors on the existence and intensity of linkages between star scientists and regional firms was reasonable and expected. Even more instructive was the finding that other reasonable variables did not play a role. Neither the international mobility background of the surveyed stars nor their age suggested significant impact. Most interestingly, however, was the result that the US factor was insignificant or even had a negative influence (see model III). This result was surprising, given the claim in the literature that the USA offers more favorable conditions for university-industry partnerships and the commercialization of (academic) research than other parts of the world (see, for instance, Mowery et al., 2001).

SUMMARY AND CONCLUSIONS

In the contemporary knowledge economy, well-educated researchers and scientists are increasingly recognized as an essential economic asset and engine for regional development. Meanwhile there exists an extensive literature on the contributions of universities to industrial innovation and their involvement in processes of regional knowledge circulation. Apart from a few notable exceptions (see, for instance, Zucker et al., 1998a; 1998b; 2002; Schiller and Revilla Diez, 2010), little attention has been given so far to knowledge-sharing activities performed by elite scientists.

In this chapter we examined to what extent and in which ways star scientists (identified by the number of citations they generated in journals in the ISI databases in the period 1981–2002) engage in regional knowledge-sharing activities. Furthermore, we intended to find out which factors determine whether star scientists engage in intraregional knowledge transfer and influence the intensity of different modes of knowledge transmission. We identified a set of mechanisms by which star scientists exchange their advanced knowledge and expertise with various actors and organizations in their regions. These included linkages to the regional academic world (academic collaborations and provision of talent for the scientific labor market) and to the policy world (advice of policy makers) as well as a differentiated typology of modes of knowledge sharing with the regional industrial world. In the latter case, we considered not only more general activities such as provision of highly qualified graduates to regional firms and fostering the entrepreneurial spirit of students, but also more specific activities like R&D collaborations with firms, selling patents to companies, and being an entrepreneur or member of a firm board.

Table 5.6 Model I: binary logit model (yes–no)

	AC_B	SL_B	IL_B	FS_B	RD_B	PA_B	EN_B	FB_B	PO_B
Returnee	1.6	0.01	0.02	0.53 *	–0.23	–0.1	0.24	0.32	0.21
Expatriate	0.47	–0.36	**–0.72** ***	–0.02	**–0.6** **	**–0.54** **	0.04	0.19	–0.02
Institution nuru	0.16	**–0.65** **	**–0.61** **	–0.4	0.32	–0.01	–0.44	–0.21	0.16
Institution corp		–1.27 *	–1.24 *	–0.36	0.98	–0.04	**1.72** **	0.86	–0.9
Institution other	–1.51	–0.32	–0.37	–0.51	0.32	**0.79** **	0.45	0.34	0.32
Type research	–0.08	0.02	**0.16** **	**0.21** ***	**0.35** ***	**0.12** **	**0.36** ***	**0.22** ***	**0.2** ***
Science group nat	0.63	–1.19 *	0	0.58	**0.87** **	**1.8** ***	–0.09	–0.33	0.43
Science group eng		–0.84	1.06 *	**1.18** **	**2.2** ***	**2.69** ***	–1.05	–0.05	0.7
Science group med	1.67	–0.64	0.13	**1.03** **	**1.29** ***	**2.57** ***	0.4	0.41	0.7*
Science group agr		–1.15	–0.86	0.79	1.21 *	1.16	–1.08	–1.26	0.24
Economic role	0.54	**0.34** **	**0.45** ***	**0.95** ***	**0.65** ***	**0.47** ***	**0.77** ***	**0.55** ***	**0.4** ***
Gender	1.09	0.23	**1.1** ***	0.09	**1.04** **	0.69	0.17	0.17	–0.07
Age	–0.3	0.06	0.07	–0.08	0.02	–0.12	0.03	–0.04	–0.08

Age_2 (10^{-3})	1.96	–0.40	0.45	0.54	0.17	0.77	–0.28	0.33	0.58
Inst. context	1.52	–0.43	–0.21	–0.31	–0.17	0.22	0.13	0.03	–0.49 *
Socialization	1.82	–0.22	–0.09	–0.34	0.05	0.13	0.02	0.3	–0.09
Early socialization	–1.46	–0.3	–0.65 *	0.22	–0.44	–0.27	–0.16	–0.29	0.35
Constant	9.61	–0.27	–4.87	–0.25	–4.8	–1.97	–6.8	–3	1.75
Obs	553	647	641	641	651	644	649	651	651
Likelihood	–35.033044	–226.56873	–331.65353	–307.22455	–286.2745	–360.01949	–227.75566	–336.81545	–358.30492
p(LR-Test)	0.4805	0.02	0	0	0	0	0	0	0
Pseudo R2	0.1625	0.064	0.1127	0.1644	0.1678	0.1192	0.1626	0.104	0.0665

* significant at the 10% level
** significant at the 5% level
*** significant at the 1% level

Table 5.7 Model II: ordered logit model (intensity)

	AC_O		SL_O		IL_O		FS_O		RD_O		PA_O		PO_O	
Returnee	0.02		–0.12		0.16		0.36	*	0.03		–0.14		0.03	
Expatriate	–0.1		–0.01		–0.31		0.35		–0.31		–0.49	*	–0.02	
Institution nuru	–0.08		–0.17		**–0.71**	***	–0.02		–0.03		–0.01		0.27	
Institution corp	0.43		–0.38		–0.82		0.31		**1.63**	***	0.21		–0.48	
Institution other	**–0.72**	**	–0.35		–0.22		–0.39		0.07		0.61	*	–0.14	
Type research	0.05		**0.1**	**	**0.19**	***	**0.19**	***	**0.31**	***	**0.12**	**	**0.19**	***
Science group nat	0.15		–0.24		–0.05		0.61	*	**0.64**	**	**1.81**	***	0.05	
Science group eng	–0.22		–0.2		**0.9**	**	**0.85**	**	**1.11**	***	**2.51**	***	0.58	
Science group med	0.08		0.1		–0.08		**0.77**	**	**0.89**	***	**2.5**	***	0.38	
Science group agr	0.4		–0.34		–0.64		0.63		0.34		1.13		0.28	
Economic role	**0.31**	***	0.08		**0.4**	***	**1.07**	***	**0.68**	***	**0.51**	***	**0.49**	***
Gender	0.35		–0.01		**0.95**	***	0		**1.02**	***	0.88	*	0.17	
Age	0.07		–0.09		0.11		–0.06		0.04		–0.08		**–0.18**	**

Age_2 (10^{-3})	–0.83	0.86	–0.85	0.43	–0.34	0.47	1.37 *
Inst. context	0.02	**–0.56** ***	–0.26	–0.07	0	0.19	–0.35*
Socialization	0.33	0.4 *	–0.18	–0.22	–0.06	0.14	–0.13
Early socialization	–0.19	–0.13	–0.14	–0.01	–0.26	–0.39	0.29
Cut1	–1.59	–3.95	5.71	1.74	5.29	3.49	–4.17
Cut2	0.54	–2.54	6.89	3.25	6.62	4.53	–3
Cut3	2.29	–0.68	8.75	4.93	8.01	6.01	–1.22
Cut4	3.78	1.19	11.07	6.52	9.48	7.86	0.06
Obs	652	647	641	641	651	644	651
Likelihood	–847.17506	–901.00406	847.6354	–844.56167	–924.59495	–582.94051	–898.62466
p(LR-Test)	0	0.0394	0	0	0	0	0
Pseudo R2	0.0288	0.0156	0.059	0.1068	0.0868	0.0792	0.0448

* significant at the 10% level
** significant at the 5% level
*** significant at the 1% level

Table 5.8 Model III: binary logit model (strong)

	AC_S		SL_S		IL_S		FS_S		RD_S		PA_S		PO_S	
Returnee	0.28		–0.53	**	0.26		0.48		0.25		–0.87		–0.07	
Expatriate	–0.13		–0.22		–0.04		0.61	*	–0.12		–0.05		0.2	
Institution nuru	0.13		0.03		–1.26	***	0.24		0.04		0.22		0.61	**
Institution corp	0.56		0.16		–0.4		0.26		3.02	***	0.55		–0.01	
Institution other	–0.49		–0.53		–0.54		–0.67		0.04		–0.48		–0.6	
Type research	0.06		0.15	**	0.26	***	0.18	**	0.26	***	0.16		0.16	**
Science group nat	0.1		0.38		0.1		0.05		–0.07		0.08		–0.77	
Science group eng	–0.42		0.09		0.91	*	–0.16		0.32		–0.25		–0.18	
Science group med	0.14		0.75	*	–0.48		0.12		0.06		–0.06		–0.33	
Science group agr	0.56		0.67		0.4		0.28		–0.67		0.51		–0.16	
Economic role	0.35	***	0.04		0.32	**	1.25	***	0.77	***	1.15	***	0.78	***
Gender	0.35		–0.39		0.54		0.71		1.19	**			0.61	
Age	0.14		–0.17		0.1		–0.1		0.04		0.09		–0.14	

Age_2 (10^{-3})	−1.44 *	1.56 *	−0.80	0.76	−0.36	−1.12	0.98
Inst. context	−0.03	−0.39	−0.71 **	0.16	−0.01	−1.11 *	−0.01
Socialization	0.42	0.59 *	−0.37	−0.43	−0.31	−0.31	−0.75 *
Early socialization	−0.49	−0.45	0.44	0.28	0.2	0.41	0.67
Constant	−4.57	3.16	−7.77	−5.58	−8.06 **	−8.51	−1
Obs	652	647	641	641	651	611	651
Likelihood	−399.78601	−332.59316	−242.23297	−245.98636	−324.82328	−102.07845	−244.57685
p(LR-Test)	0.0001	0.0219	0	0	0	0.0038	0
Pseudo R2	0.0568	0.0441	0.1051	0.176	0.1405	0.147	0.1076

* significant at the 10% level
** significant at the 5% level
*** significant at the 1% level

Drawing on the results of a worldwide survey of 720 star scientists in 21 different areas of research we have shown that these highly cited individuals are rather strongly embedded in their regions. The surveyed members of the scientific elite had a rather positive attitude towards contributing to regional development and they exhibited various connections to other actors and organizations in their region. Investigating the relative importance of various types of intraregional knowledge transfer, we highlighted that academic collaboration within the region tended to be almost ubiquitous, closely followed in level by providing talent for the scientific labor market. However, the role of the surveyed star scientists was not restricted to these academic core activities. We also observed that linkages to regional firms and policy makers were rather common. We found evidence that more than two thirds of the investigated stars acted as a source of highly qualified graduates for regional companies and fostered entrepreneurial spirit of their students. Similar proportions were also found to engage in more specific activities like performing R&D projects with firms and providing policy advice. But even rather specific activities like selling patents to firms and being a regional entrepreneur or a board member were reported by a substantial proportion of the top researchers included in our sample. Examining the intensity by which the surveyed star scientists adopted different modes of knowledge sharing confirms our conclusion that these stars were by no means detached inhabitants of the academic ivory tower. On the contrary, the evidence suggests that they are strongly embedded in their regional economies.

Finally, using three multivariate models we also investigated factors that influence the existence and intensity of different modes of regional knowledge transfer adopted by star scientists. However, only a few variables turned out to have a positive, statistically significant impact on such activities. These comprised the general attitude about the role of researchers in regional development and the degree of applied research. The other factors included in our models (such as the international mobility background of top researchers, their age, and surprisingly the US factor) did not play a role. Obviously, other variables than those which were considered in our models may have a stronger explanatory power. Thus, it remains obscure who the top scientists are who engage in intraregional knowledge-sharing activities. To unravel this issue is a key question for further research.

REFERENCES

Agrawal, A. and R. Henderson (2002), 'Putting patents in context: exploring knowledge transfer from MIT', *Management Science*, **48**, 44–60.

Azoulay, P., W. Ding and T. Stuart (2009), 'The impact of academic patenting on the rate, quality and direction of (public) research output', *Journal of Industrial Economics*, **57**, 637–76.

Baba, Y., N. Shichijo and S. Sedita (2009), 'How do collaborations with universities affect firms' innovative performance? The role of "Pasteur scientists" in the advanced materials field', *Research Policy*, **38**, 756–64.

Bergman, E. (2010), 'Knowledge links between European universities and firms: a review', *Papers in Regional Science*, **89**, 311–33.

Breschi, S., F. Lissoni and F. Montobbio (2007), 'The scientific productivity of academic inventors: new evidence from Italian data', *The Economics of Innovation and New Technology*, **16**, 101–18.

Calderini, M., C. Franzoni and A. Vezzulli (2007), 'If star scientists do not patent: the effect of productivity, basicness and impact on the decision to patent in the academic world', *Research Policy*, **36**, 303–19.

Etzkowitz, H. and L. Leydesdorff (2000), 'The dynamics of innovation: from national systems and "Mode 2" to a triple helix of university – industry – government relations', *Research Policy*, **29**, 109–123.

Feldman, M. and R. Owens (2007), 'Bringing science to life. Introduction to the special issue', *Journal of Technology Transfer*, **32**, 127–30.

Furukawa, G. and A. Goto (2006), 'Core scientists and innovation in Japanese electronics companies', *Scientometrics*, **68**, 227–40.

Gunasekara, C. (2006), 'Reframing the role of universities in the development of regional innovation systems', *The Journal of Technology Transfer*, **31**, 101–13.

Horowitz, I. (1966), 'Some aspects of the effects of the regional distribution of scientific talent on regional economic activity', *Management Science*, **13**, 217–32.

Keeble, D. (2000), 'Collective learning processes in European high-technology milieux', in D. Keeble and F. Wilkinson (eds), *High-Technology Clusters, Networking and Collective Learning in Europe*, Aldershot: Ashgate, pp. 199–229.

Laudel, G. (2005), 'Migration currents among the scientific elite', *Minerva*, **43**, 377–95.

Lowe, R. and C. Gonzales-Brambila (2007), 'Faculty entrepreneurs and research productivity', *Journal of Technology Transfer*, **32**, 173–94.

Mowery, D. and B. Sampat (2005), 'Universities in national innovation systems', in J. Fagerberg, D. Mowery and R. Nelson (eds), *The Oxford Handbook of Innovation*, Oxford: Oxford University Press, pp. 209–39.

Mowery, D., R. Nelson, B. Sampat and A. Ziedonis (2001), 'The growth of patenting and licensing by US universities: an assessment of the effects of the Bayh-Dole Act of 1980', *Research Policy*, **30**, 99–119.

OECD (2006), *Revised Field of Science and Technology (FOS) Classification in the Frascati Manual*, Paris: OECD.

Pavitt, K. (1984), 'Sectoral patterns of technical change: towards a taxonomy and a theory', *Research Policy*, **13**, 343–73.

Schiller, D. and J. Revilla Diez (2010), 'Local embeddedness of knowledge spillover agents: empirical evidence from German star scientists', *Papers in Regional Science*, **89**, 275–94.

Stephan, P., S. Gurmu, A. Sumell and G. Black (2007), 'Who's patenting in universities? Evidence from the survey of doctorate recipients', *The Economics of Innovation and New Technology*, **16**, 71–99.

Thorn, K. and L. Holm-Nielsen (2008), 'International mobility of researchers and scientists: policy options for turning a drain into a gain', in A. Solimano (ed.), *The International Mobility of Talent*, Oxford: Oxford University Press, pp. 145–67.

Tödtling, F., P. Lehner and M. Trippl (2006), 'Innovation in knowledge intensive industries: the nature and geography of knowledge links', *European Planning Studies*, **14**, 1035–58.

Trippl, M. (2011a), 'Scientific mobility and knowledge transfer at the interregional and intraregional level', First published on: 11 March 2011 (iFirst), *Regional Studies*, 1–15.

Trippl, M. (2011b), 'Islands of innovation as magnetic centres of star scientists? Empirical evidence on spatial concentration and mobility patterns', First published on: 17 March 2011 (iFirst), *Regional Studies*, 1–16.

Trippl, M. and G. Maier (2011a), 'Knowledge spillover agents and regional development', in P. Nijkamp and I. Siedschlag (eds), *Innovation Growth and Competitiveness. Dynamic Regions in the Knowledge-based World Economy*, Berlin, Heidelberg: Springer, pp. 91–111.

Trippl, M. and G. Maier (2011b), 'Star scientists as drivers of the development of regions', in P. Nijkamp and I. Siedschlag (eds), *Innovation Growth and Competitiveness. Dynamic Regions in the Knowledge-based World Economy*, Berlin, Heidelberg: Springer, pp. 113–34.

Van Looy, B., M. Ranga, J. Callaert, K. Debackere and E. Zimmermann (2004), 'Combining entrepreneurial and scientific performance in academia: towards a compounded and reciprocal Matthew-Effect', *Research Policy*, **33**, 425–41.

Zucker, L. and M. Darby (1996), 'Star scientists and institutional transformation: invention and innovation in the formation of the biotechnology industry', *Proceedings National Academy of Sciences*, **93**, 12709–16.

Zucker, L. and M. Darby (2001), 'Capturing technological opportunity via Japan's star scientists: evidence from Japanese firms' biotech patents and products', *Journal of Technology Transfer*, **26**, 37–58.

Zucker, L. and M. Darby (2006), 'Movement of star scientists and engineers and high-tech firm entry', *NBER Working Paper* No. 12172, April 2006.

Zucker, L., M. Darby and J. Armstrong (1998a), 'Geographically localized knowledge: spillovers or markets', *Economic Inquiry*, **36**, 65–86.

Zucker, L., M. Darby and M. Brewer (1998b), 'Intellectual human capital and the birth of U.S. biotechnology enterprises', *American Economic Review*, **88**, 290–306.

Zucker, L., M. Darby and J. Armstrong (2002), 'Commercializing knowledge: university science, knowledge capture, and firm performance in biotechnology', *Management Science*, **48**, 138–53.

6. Firm productivity in innovative urban milieus

Frank G. van Oort and Otto Raspe

INTRODUCTION

Due to the substantial theoretical foundation of the role of knowledge in modern growth theory, knowledge is regarded as an explicit and crucial factor for generating sustained economic growth in Western economies (Audretsch et al., 2006). Within this theorizing, knowledge spillovers are considered a key element in these new growth theory models and form a mechanism in firm-external economies (Koo, 2005). According to this view, individual firms produce (technological) knowledge. At first, this is private to the firm; afterwards, it might spill over to the rest of the economy as it can be copied at almost no cost by other firms. It even might become social knowledge, acting as an external effect in enhancing the productivity of all firms. With the spillover effect, an aggregate production function with otherwise constant or decreasing returns to scale, may exhibit increasing returns to scale, allowing sustained long-run growth. An implication of this view is that a firm, not able to innovate on its own, can benefit from the research findings of firms working along similar lines (Sena, 2004).

The geographical and regional economics literature on knowledge spillovers confronts us with the fact that despite its public good properties, knowledge does not diffuse instantaneously to production facilities around the world (Acs and Plummer, 2005; Döring and Schnellenbach, 2006). In this literature there is a tradition in analyzing the local advantages of proximity or agglomeration, questioning whether regional economic growth is higher in regions where more organizations or knowledge are concentrated. Also, the territorial innovation literature suggests that learning processes take place at the local and regional level, and that this is crucial for the creation and acceptance of innovation. Arguments of proximity, face-to-face interaction, knowledge tacitness, the stickiness of information, long-term trust-based relationship between firms, labor market mobility, and spin-offs all point to advantages of regions and cities in explaining economic growth due to knowledge spillovers (McCann and Simonen, 2005).

While the empirical evidence about the linkages between agglomerations and growth focuses at regional and local analysis, the relationship should actually and most profoundly hold at the micro or firm level. But little is known about the locational impact on firm performance, as measured in terms of individual performance (Raspe and Van Oort, 2008). The firm and the geographical cluster in innovation and learning processes have typically been studied separately (Mariani, 2004), and only a few attempts have been made to relate a firm's innovativeness and performance to regional variables, so as to provide a clear distinction between firm- and region-specific effects (Beugelsdijk, 2007). Audretsch and Dohse (2007) and Raspe and Van Oort (2008) indicate that the reasons for this omission are both conceptual and empirical in nature. At the conceptual level, there only a few models that link the performance of individual firms to regional (knowledge and human capital) characteristics. At the empirical level, analyzing firm growth (in a spatial context) requires longitudinal data at the establishment or enterprise level, which are often not available.

Applying multilevel models, in this chapter we focus on the relation between firm level productivity and firms' embedding in innovative and knowledge-rich urban environments in the Netherlands. We find that it is important to take firm-specific characteristics simultaneously with region-specific characteristics into account, since a large part of what traditionally is assigned to the impact of the region should actually be assigned to firm-specific characteristics. We conclude that innovative urban environments are profitable for firm level productivity – an outcome that confirms but also nuances previous empirical findings on the spatial impact of the knowledge economy. The presented insights are important for policy makers proposing economic development by geographically focused policy investments.

The chapter is further organized as follows. The next section describes the conceptual framework of our Dutch research on firm productivity embedded in urban knowledge densities. The third section presents our empirical approach, followed by the results of the multilevel models in the fourth section. We conclude in the final section.

FIRM-SPECIFIC CHARACTERISTICS AND SPATIALLY BOUNDED KNOWLEDGE EXTERNALITIES

One of the leading indicators in the debate of locational impacts on firm performance is productivity (Ciccone and Hall, 1996). In particular, studies on industrial dynamics and entrepreneurship have a tradition in theorizing why firm productivity varies systematically across firms and which internal factors are related to this phenomenon. We do not give a full overview of this debate (see

Raspe and Van Oort, 2006; 2008 for that), but rather distillate the most relevant elements that are related to productivity.

Firm-specific Characteristics

Knowledge creation and entrepreneurial learning are stressed as important strategic activities of the firm, and spillovers of knowledge are crucial in generating innovative output (Parker, 2005). Able entrepreneurs survive and grow, while the less able (or unlucky) exit the market. New scientific and technological knowledge is often an important source of entrepreneurial opportunities. Acs and Armington (2004) introduce the entrepreneur as a conduit for transforming new knowledge into new economically valuable business opportunities. A main mechanism for recognizing new opportunities and actually implementing them by starting new economic activities involves knowledge spillovers (Audretsch et al., 2006). Entrepreneurship can take shape by new firm formation, start-ups and spin-offs, but also by incorporating business opportunities in incumbent firms. Firm size largely determines a firm's resource base, competencies and scale advantages. For instance, due to internal economies of scale causing a reduction in per-unit costs over the number of units produced, efficiency advantages emerge from larger firm sizes (Carroll and Hannan, 2000). Small firms, due to their limited initial size, have to overcome these disadvantages, unlike larger firms.

A number of studies bring to the fore that in addition to size, the ossification of routines and learning processes is important for firms' productivity (Sutton, 1997). It is especially important to take a firm's age into account (Jovanovic, 1982). Age is the representative of the selection processes due to learning over time. This leads to the assumption that older firms function more efficiently, indicating that firms can learn from their own experience and successful firms grow and become more efficient (i.e., reduce their costs), while the unsuccessful ones face lower productivity levels and even may exit the sector. On the other hand, older firms may suffer from path dependency and see themselves as confronted with 'lock-in', meaning that older firms have lower productivity rates. This lock-in can be described as the competency trap; becoming quite good at doing any one thing reduces an organization's capacity to absorb new ideas and to do other things.

In addition to size and age, the type of economic activity (sectors) is important for firm performance. Industry-related effects capture various technology and knowledge dimensions such as technological opportunity, appropriability regimes, or the emergence of dominant designs along the technology life cycle (Breschi et al., 2000). This results in productivity differences stemming from sector membership.

The management science literature has recently stressed the importance of firm competences and resources, especially related to a firm's internal knowledge base as enhancing productivity. The ability to identify, assimilate and apply for commercial purposes know-how generated outside a firm's own organization is one of the most relevant business characteristics in determining the productivity and innovativeness of firms (Cohen and Levinthal, 1989). The so-called competence-based approach emphasizes the importance for performance of largely tacit and socially produced and reproduced knowledge. In particular, the information used by innovators has a tacit component; there is information that is difficult to set down in blueprints or to codify completely, and hence it is difficult to communicate at a distance (Glaeser, 1999). Attention is also paid to the use of information and communication technologies (ICT). Drennan (2002) analyses the potentials related to an increased accessibility of codified information through the adoption of ICTs. Recent studies suggest that ICT as a general-purpose technology enhances productivity by making working processes more efficient (Becchetti et al., 2003).

Spatially Bounded Knowledge Externalities[1]

Firms cluster and profit from agglomerated densities to economize on the transport of goods, people and ideas. In particular, knowledge spillovers (unpaid externalities related to information and knowledge) form a mechanism in firm-external economies and are key elements of productivity and growth (Henderson, 2007). Cities then are considered as centers of innovation, technological progress, ideas and knowledge.

These regional or urban externalities are also known as agglomeration economies. These can be divided into localization economies, urbanization economies and Jacobs' externalities (Frenken et al., 2007). Localization economies are externalities available to all local firms within the same sector. They usually take the form of what are called Marshallian (technical) externalities, whereby the productivity of labor in a given sector in a given city increases with total employment in that sector. Urbanization economies are available to all local firms, irrespective of sector. They arise from urban size and density. Relatively more populous localities are also more likely to house universities, industry research laboratories, trade associations and other knowledge-generating organizations, and usually are better accessible. So-called Jacobs' externalities are external economies stemming from a local variety of sectors. A diverse urban industry mix improves the opportunities to interact, copy, modify and recombine ideas, practices and technologies across industries (Jacobs, 1969). In the literature, urbanization economies and Jacobs' externalities are often conceptualized equal, which is not correct.

Raspe and Van Oort (2006) show that the recent interest in the knowledge economy is embedded in a long tradition. The knowledge economy is usually understood as an economy in which the production factors labor and capital are aimed at the development and application of new technologies (OECD, 1996). This definition falls short in the sense that the ultimate goal of the knowledge economy is taken to be the application of new technologies as such, while, in fact, this application is instrumental to the goals of innovation and economic (productivity) growth. Since its introduction, many theoretical and empirical contributions have therefore refined and broadened the concept. In line with Raspe and Van Oort (2006) we distilled (measurable) indicators from this literature that are relevant for firm level and contextual growth models.

Knowledge is defined as the ability to recognize and solve problems, by collecting, selecting and interpreting relevant information. Hence, a basic feature of the knowledge economy is the use of knowledge in interrelationships among market actors to produce goods and services, from the first idea to final products. A well-educated workforce has ample opportunities to absorb and use information. In measuring the localized knowledge economy, we therefore use the average educational level of the working population per municipality as a first indicator. Florida (2002) though, identifies creative capital embodied in knowledge workers and artists as a major indicator of the knowledge economy. The difference between human and creative capital is that the 'creative class' (as Florida labels it) does not necessarily need to have a high educational level in order to create added value. In addition to direct productivity effects produced by knowledge workers, Florida emphasizes indirect growth effects from consumption by the creative class in the amenity-rich urban environments they live in. Since data on the creative class itself is not available, we use a proxy, that is, the density of creative industries, as a second knowledge economy indicator. The literature on the knowledge economy also emphasizes two indicators that reflect accessibility and transfer of knowledge. Particularly, Drennan (2002) analyses the growth potentials of firms related to an increased accessibility of information through the adoption of information and communication technologies (ICT). Hence, we take ICT density (measured by computer usage per employee per industry) as a third indicator. Cooke and Morgan (1998) identify social, cultural and communicative capital as sources of employment growth. We measure this variable via the classification of occupations according to the degree of communicative skills needed for interaction. We define a sectorally weighted average degree of communication skills as a fourth indicator.

Our definition of the locally defined knowledge economy also addresses technical and production oriented aspects. Traditionally most attention has been paid to research and development (R&D). We use the sectorally weighted share of R&D employees as a fifth indicator. Additionally, Cortright and Mayer (2001) emphasize the role of high- and medium-tech firms as indicators of the knowledge economy and drivers of economic and employment growth. Besides R&D intensity, the OECD (1996) argues that high- and medium-tech firms are characterized by their export intensity. We take the density of high- and medium-tech industries as sectorally defined by the OECD relative to the total population of firms, as a sixth indicator. Finally, innovation is generally regarded as the most important driver of economic and employment growth. Several indicators of innovation exist, like new product announcements, publications, patents and firm self-ratings. In this chapter we use firm self-ratings in terms of new products and processes (as expressed by firms in the Third Community Innovation Survey for the Netherlands). We distinguish between technical and non-technical innovations. While technical innovations relate to new products and production processes, non-technical innovations concern management, organization and services. Both aspects are taken into account as the proportion of innovative firms in a municipality. They are our seventh and eighth indicators.

These variables are indicators of underlying latent variables, and are therefore strongly correlated. Regions specialized in ICT-intensive activities usually also are characterized by a highly educated labor force. And R&D-intensive regions usually also contain many high- and medium-tech firms. Direct inclusion of the indicators would thus lead to multicollinearity and hence an increase of the estimated variances of the estimators of their coefficients so that one might be led to drop some of the variables incorrectly from the productivity model. Therefore, we include three latent variables into the model rather than the indicators (Raspe and Van Oort, 2006). The latent variables are related to their observable indicators via a (principal component) measurement model. We distinguish the following latent variables:

1. 'knowledge workers' with indicators: ICT sensitivity, educational level, creative class and communicative skills;
2. 'R&D' with indicators: the density of high- and medium-tech firms and the share of R&D employees;
3. 'innovativeness' with indicators: technical and non-technical innovations.

The factor scores for these principal components are shown in Table 6.1. Bold values mark the loading of the indicators that are taken together in the three factors.

To finalize the literature review, some more conventional variables should also be taken into account to control for more general spatial advantages (see Frenken et al., 2007; Audretsch et al., 2006). These are: (1) general regional growth, implying increasing market size, (2) spatial investment levels, an indication of physical renewals in the production environment, (3) sectoral wages, (4) newly built business premises, since they attract more than average economic activity that previously was not present in that location, (5) physical accessibility and (6) the presence of a university.

Table 6.1 Factor scores of principal components

Indicators	Factor 1 'Knowledge workers'	Factor 2 'Innovation'	Factor 3 'R&D'
ICT-sensitivity	**0.753**	0.365	0.268
Education level	**0.949**	0.164	0.044
Creative economy	**0.516**	0.024	–0.198
Communicative skills	**0.927**	0.040	–0.069
High-tech and medium-tech	–0.175	0.146	**0.840**
Research and development	0.080	0.129	**0.836**
Innovation (technological)	0.130	**0.878**	0.246
Innovation (non-technological)	0.147	**0.914**	0.054

Source: Raspe and Van Oort (2006)

EMPIRICAL APPROACH: VARIABLES AND MODELING

We model firm productivity, simultaneously taking both firm-specific and region-specific independent variables into account. Therefore we use multi-level models (see Appendix 6.1). For the firm-level data, our chapter draws on a survey with data concerning 2009 firms in manufacturing and business services in the Netherlands. The survey was conducted in 2005 (Raspe and Van Oort, 2008). At the firm level, our performance indicator is the level of labor productivity ($PROD_{ij}$), measured as the added value of a firm per employee. The firm's added value is determined as the yearly turnover minus purchases (all intermediate goods and services needed in the production process of the firm). The added value includes the firm's taxes, subsidies, wages and profits. Labor productivity is distilled by dividing the added value at the firm level by the number of employees.

We take firm size ($SIZE_{ij}$) into account by the firm's turnover. Our hypothesis is that firm size is positively related to labor productivity; larger firms are better organized and potentially profit from divisions of labor within the firm. Age (AGE_{ij}) is measured by the number of years since the firm has entered business (year of foundation of the establishment). Our hypothesis is that age is positively related to a firm's productivity, unless firms suffer from lock-in (represented by including a squared variable AGE_SQ_{ij} simultaneously with the age variable). A firm's absorptive capacity depends to a large extent on its existing (prior) stock of knowledge, much of which is embedded in its products, processes and people (Cohen and Levinthal, 1989). Basic R&D capabilities in general constitute a firm's absorptive capacity (Cassiman and Veugelers, 2006). For business services firms, which conduct less R&D but rely heavily on professional knowledge, the number of knowledge-intensive jobs has a similar meaning (Illeris, 1996). We measure a firm's total number of knowledge-intensive occupations as a percentage of the total number of jobs ($KI.OC_{ij}$). This includes for manufacturing firms, the number of occupations in research and development and for business services, the number of occupations in consulting (marketing- and design-related). Other jobs within the firm concern management, facilitating services, purchasing, logistics or production. We account for the use of face-to-face ($F2F_{ij}$) and information and communication technology (ICT_{ij}) contact by selecting these transfer mechanisms from a survey question on the types of transfer mechanisms used in inter-firm relationships. They are expressed as the sum of all types of transfer mechanisms (personal, written, telephone, e-mail, e-commerce and other types of contact). These variables are direct measurements of how a firm handles its inter-firm relationships by different communication channels. $F2F_{ij}$ is defined as the degree to which it does so by personal contacts, ICT_{ij} by e-mail plus e-commerce contacts.

The external agglomeration sources that we take into account are localization, urbanization and Jacobs' externalities, as well as the three regional knowledge economy factors introduced in the previous section. Localization economies are measured by specialization measures for four types of economic sectors: labor-, capital- and knowledge-intensive manufacturing and business services (respectively $SPEC.AI_j$, $SPEC.CI_j$, $SPEC.KI_j$ and $SPEC.KIS_j$). A density indicator measures urbanization economies: the number of jobs per square kilometer ($DENSITY_j$). Jacobs' externalities are measured by a diversity measure (Gini coefficient; $GINI_j$) that takes into account the employment distribution across 58 detailed sectors in the region. Regions can vary according to whether employment is strongly concentrated in one sector or distributed equally among sectors (relative to the national distribution). Since direct technological-oriented spillovers and human capital are considered as key externalities for productivity and growth, we empirically focus on the knowledge externalities 'research and development', 'innovation' and 'knowledge workers'.

In addition to the knowledge-related determinants, we control in our models for general regional growth ($REG.GROWTH_j$) measured by the relative employment growth in the region over the last six years. Spatial investments levels ($INVEST_j$) are measured by investments in immobile capital goods, excluding houses. Sectoral differences in wages ($WAGE_j$) are measured over regions. Newly built business premises ($BUS.SITES_j$) are measured by the growth in business sites in hectares (see also Van Oort, 2004). The number of jobs that can be reached by road within 45 minutes' travel time measures the degree of physical accessibility ($ACCESAB_j$) of a region. Finally, a dummy for the presence of a university is included.

Where we control for sectoral fixed effects, we include three industry dummies: AI_{ij}, CI_{ij}, KI_{ij}, respectively representing industry memberships in labor-intensive manufacturing, capital-intensive manufacturing and knowledge-intensive manufacturing. We use knowledge-intensive business services as a reference. All spatial variables (reflecting spatial externalities) draw upon a database that includes all establishments in the Netherlands involved in all types of economic activities (LISA database). This database is aggregated to an urban level (128 municipalities, corresponding to cities). In the models, we include a lagged effect in the relation between the firm's knowledge context and the firm's performance. Henderson (2003) shows that it takes time for knowledge to spill over and get embedded in firms. We therefore use a three-years' lag by linking firm productivity in 2005 to variables in the firm's sectoral environment in 2002.

EMPIRICAL RESULTS

It is important to distinguish the firm-level-compositional source of variation from the regional-contextual ones. The null model (indicating the earlier mentioned intra-class correlation) gives an indication of the part of the variation that can be assigned to regional-specific characteristics. Regression results of alternative models explaining firm level productivity are summarized in Table 6.2. Note from this table that 2.3% of firm productivity can be related to its location (and more that 97% to matters of internal organization). Our first conclusion therefore is that the region plays a role but that its impact is minor compared to firm internal factors.

Further, Table 6.2 presents the models for firm level productivity including independent variables on both the firm and the regional level. Model 1 shows the results when firm-specific characteristics are controlled for. Firm size is significant and positively related to productivity. The larger the firm, the more economies of scale influence firm productivity. These results are in line with many findings in the literature (Caroll and Hannan, 2000). Model 1 also tests for the age of a firm. Besides age, an age-quadratic term is simultaneously included as a proxy for possible lock-in of the oldest firms. Model 1 shows that age is significant and positively attached to productivity levels, while age-squared is significant and negatively attached, confirming our hypothesis. Model 1 also provides insight on the effect of knowledge-based firm-level variables, like the share of knowledge-intensive occupations, the

Table 6.2 Regression results of alternative models explaining firm level productivity

	Null	Model 1	Model 2	Model 3	Model 4
Firm					
Constant	10.627***	3.793**	10.081**	3.170***	4.055
	(0.084)	(0.741)	(0.642)	(0.969)	(2.500)
SIZE		0.423***	–	0.414***	0.415***
		(0.034)		(0.034)	(0.034)
AGE		1.434*	–	1.544*	1.548**
		(0.869)		(0.868)	(0.869)
AGE_SQ		–0.591*	–	–0.631**	–0.634**
		(0.323)		(0.322)	(0.323)
KI.OC		0.003	–	0.004	0.005
		(0.018)		(0.018)	(0.018)
F2F		0.029	–	0.033	0.034
		(0.027)		(0.027)	(0.027)
ICT		0.066**	–	0.066**	0.066**
		(0.029)		(0.029)	(0.029)
Regional					
SPEC.AI			0.158	0.139	0.102
			(0.129)	(0.124)	(0.131)
SPEC.CI			–0.090	–0.105	–0.090
			(0.084)	(0.081)	(0.086)
SPEC.KI			0.033	–0.016	–0.007
			(0.083)	(0.079)	(0.081)
DENSITY			0.148**	0.135**	0.119*
			(0.072)	(0.069)	(0.071)
GINI			–0.097	0.025	0.144
			(0.346)	(0.329)	(0.307)
'KW'			–0.160*	–0.132	–0.135
			(0.093)	(0.091)	(0.091)
'R&D'			0.096	0.116	0.135
			(0.084)	(0.081)	(0.082)
'INN'			0.306***	0.205**	0.204**
			(0.079)	(0.076)	(0.100)
REG. GROWTH					0.357
					(0.896)

	Null	Model 1	Model 2	Model 3	Model 4
INVEST					0.007 (0.061)
WAGE					–0.112 (0.093)
BUS.SITES					–0.008 (0.072)
ACCESAB					–0.081 (0.169)
Sectoral fixed effects	No	Yes	No	Yes	Yes
–2*loglikelih.	9870.6	9697.9	9855.1	9687.7	9683.4
R square *i*	–	7.7%	–	7.7%	7.7%
j		48.0%	64.3%	79.5%	99%
Intraclass correl.	2.3%				

Notes:
For explanations of the variables abbreviations see p. 107.
Standard deviation between parentheses, *** significant at the 0.01 level, ** significant at the 0.05 level, * significant at the 0.10 level.
Robustness checks: industry, agglomeration, time. In our models, we tested for specifications of industry, agglomeration and time-fixed effects. We tested for industry fixed effects, making a distinction in labor-, capital- and knowledge-intensive manufacturing and business services. Dummy coefficients are not included in the table.
Variance explained knowing that there is a contextual effect for firm's productivity and that firm internal characteristics are dominant, the question remains: 'how much variance is explained?' In multilevel regression analysis, the issue of modeled or explained variance as test statistic is complex. It can be estimated by examining the residual error variances in a sequence of models where the null model is the baseline model. This must be issued level by level. For our models, we see that the proportion of variance explained at the firm level is 7.7%. Model 2a shows that although no level 2 variables are added, 48% of the municipal-level variance is explained by the firm's internal characteristics. If the proportion of, for example, size and sector are not exactly the same in all regions, the regions differ in the average size or sectors of their firms; this variation can explain some of the variance in average productivity between regions. In model 3a, we see that adding regional knowledge factors raises the *R* squared by almost 15% points (to 79.5%).

share of face-to-face interaction and ICT usage. We only find a positive effect of ICT usage. The more ICT firms use their inter-firm relationships, the more productive firms are. Our variables, reflecting absorptive capacity and face-to-face interaction, are not significant. The size variable might capture part of the effect of the absorptive capacity indicator – to avoid multicolliniarity problems we do not test them simultaneously. While face-to-face interaction is assumed to be important for internalizing tacit knowledge, we do not find that the use of more of these interactions leads to better firm performance.

When we focus on the three spatial knowledge indictors (R&D, innovation and education) we learn that firm productivity can indeed be related to three kinds of external knowledge sources (model 2). Locally bounded innovation especially enhances firm productivity. This means that firms profit from being located in an average innovative region. Remarkably, our analyses show that a high intensity of external localized research and development does not enhance firm's productivity. Apparently the regional introduction of new products and processes is more important for firms than regional R&D. Also, the human capital variable on 'knowledge workers' is not significant, meaning that firm productivity is not influenced by the regional amount of highly educated, creative and communicative workers. An explanation for this might be that the highest intensities of this variable co-locate with the urbanization (density) variable. Although the correlation between those variables is not too high, it can be argued that density captures part of the knowledge workers' variable. The urbanization variable is positively significant, meaning that firms are more productive in the most dense and agglomerated part of the landscape.

Model 3 presents a full model in which firm-specific and regional-specific characteristics are simultaneously determined. This specification shows that while controlling for firm-specific characteristics, the spatial effects become less important. We did not control for these firm characteristics in models 1–2 in Table 6.2, and assigned a considerable part of the regional variation to spatial innovation spillovers or externalities. This effect disappears due to unequal distribution – especially that of firm size – over space. We conclude that it is indeed important to control for firm level heterogeneity to reduce the potential upward biases and possible exaggerations of the spatial impact. Model 4 finally controls for the more traditional localization, urbanization and conventional regional variables that can enhance firm productivity. These variables are generally not significant in our models.

CONCLUSIONS

In this chapter we contribute to discussion on linking firm external knowledge environments to individual firm productivity. We question whether knowledge characteristics of a firm's location impact its growth. We defined the spatial knowledge environment of firms on the level of Dutch cities as a natural context to understand the mechanics of productivity growth in relation to the compact geographical nature of communication and knowledge flows. When knowledge is not easily exchanged over distance, firms tend to locate in proximity to capitalize on the knowledge stock in neighboring firms and institutions. Based on the knowledge economy literature, we not only focus on R&D or technological externalities, but we value complementary indicators like the successful introduction of new products and services to the market (innovation) and indicators of skills of employees (knowledge workers) as well. Simultaneously controlling for the firm's age, size and sectoral membership, we indeed find indications that firms experience higher productivity when located in a region with a higher intensity of successful innovative firms. R&D density and knowledge workers (human capital) density has no significant impact on firm productivity.

This paper empirically shows the significance of controlling for firm-specific characteristics while simultaneously testing for spatial knowledge-economic effects. This follows up on recent critical authors who claim that while the empirical evidence about the linkages between regionally bounded advantages and economic growth are focused on the regional and local levels, the relationship should actually and most profoundly hold at the micro or firm level. We found that the region contributes almost 2.5% to productivity at the firm level. Analyzing what kinds of externalities contribute to this spatial impact, we found that external innovation sources enhance individual firm productivity. This effect turns out to be a composition effect due to firm-specific characteristics, mainly firm size and type of industrial sectors.

The presented insights are important for policy makers proposing economic development by geographically focused policy investments. Heterogeneous processes at the firm and regional levels may not align with those of regional planners, who hope to develop innovative and high-performance regions in general. Regional economic strategies should take into account these industrial organization aspects, since the firm-context interaction turns out to be a complex one depending on both firm-specific and region-specific characteristics. We also recommend that regional innovation policies broaden their perspective from R&D and human capital ('knowledge workers') – indicators policy makers can more easily influence – to actual innovative successful firms.

NOTE

1 This section is based on Raspe and Van Oort (2006, 2008).

REFERENCES

Acs, Z.J. and C. Armington (2004), 'Employment growth and entrepreneurial activity in cities', *Regional Studies*, **38**, 911–27.

Acs, Z.J. and L.A. Plummer (2005), 'Penetrating the "knowledge filter" in regional economies', *The Annals of Regional Science*, **39**, 439–56.

Audretsch, D.B. and D. Dose (2007), 'Location: a neglected determinant of firm growth', *Review of World Economics*, **143**, 79–107.
Audretsch, D.B., M.C. Keilbach and E.E. Lehmann (2006), *Entrepreneurship and Economic Growth*, Oxford: Oxford University Press.
Becchetti, L., D. Bedoya and L. Paganetto (2003), 'ICT investment, productivity and efficiency: evidence at firm level using a stochastic frontier approach', *Journal of Productivity Analysis*, **20**, 143–67.
Beugelsdijk, S. (2007), 'The regional environment and a firm's innovative performance: a plea for a multilevel interactionist approach', *Economic Geography*, **83**, 181–99.
Breschi, S., F. Malerba and L. Orsenigo (2000), 'Technological regimes and Schumpeterian patterns of innovation', *The Economic Journal*, **110**, 388–410.
Caroll, G.R. and M.T. Hannan (2000), *The Demography of Corporations and Industries*, Princeton: Princeton University Press.
Cassiman, B. and R. Veugelers (2006), 'In search of complementarity in innovation strategy: Internal R&D and external knowledge acquisition', *Management Science*, **52**, 68–82.
Ciccone, A. and R.E. Hall (1996), 'Productivity and the density of economic activity', *American Economic Review*, **86** (1), 54–70.
Cohen, W.M. and D.A. Levinthal (1989), 'Innovation and learning: the two faces of R&D', *The Economic Journal*, **99**, 569–98.
Cooke, P. and K. Morgan (1998), *The Associational Economy. Firms, Regions and Innovation*, Oxford: Oxford University Press.
Cortright, J. and H. Mayer (2001), 'High-tech specialization: a comparison of high-tech centers', *The Brookings Survey Papers*, January 2001, 1–18.
Döring, T. and J. Schnellenbach (2006), 'What do we know about geographical knowledge spillovers and regional growth?: a survey of the literature', *Regional Studies*, **40**, 375–95.
Drennan, M.P. (2002), *The Information Economy and American Cities*, Baltimore, MD: The Johns Hopkins University Press.
Florida, R. (2002), *The Rise of the Creative Class*, New York: Basic Books.
Frenken, K., F.G. van Oort and T. Verburg (2007), 'Related variety, unrelated variety and economic growth', *Regional Studies*, **41**, 685–97.
Glaeser, E.L. (1999), 'Learning in cities', *Journal of Urban Economics*, **46**, 254–77.
Henderson, J.V. (2003), 'Marshall's scale economies', *Journal of Urban Economics*, **53**, 1–28.
Henderson, J.V. (2007), 'Understanding knowledge spillovers', *Regional Science and Urban Economics*, **37**, 497–508.
Illeris, S. (1996), *The Service Economy: A Geographical Approach*, Chichester: John Wiley & Sons.
Jacobs, J. (1969), *The Economy of Cities*, London: Jonathan Cape.
Jovanovic, B. (1982), 'Selection and evolution of industry', *Econometrica*, **50**, 649–70.
Koo, J. (2005), 'Agglomeration and spillovers in a simultaneous framework', *The Annals of Regional Science*, **39**, 35–37.
Mariani, M. (2004), 'What determined technological hits? Geography versus firm competencies', *Research Policy*, **33**, 1565–82.
McCann, P. and J. Simonen (2005), 'Innovation, knowledge spillovers and local labor markets', *Papers in Regional Science*, **84**, 465–85.
OECD (Organisation for Economic Co-operation and Development) (1996), *The Knowledge-based Economy*, Paris, OECD.

Parker, S.C. (2005), 'The economics of entrepreneurship: what we know and what we don't', *Foundations and Trends in Technology, Information and Operations Management*, **1**, 1–54.
Raspe, O. and F.G. van Oort (2006), 'The knowledge economy and urban economic growth', *European Planning Studies*, **14**, 1209–34.
Raspe, O. and F.G. van Oort (2008), 'Firm growth and localized knowledge externalities', *Journal of Regional Analysis and Policy*, **38**, 100–116.
Sena, V. (2004), 'Total factor productivity and the spillover hypothesis: some new evidence', *Int. J. Production Economics*, **92**, 31–42.
Sutton, J. (1997), 'Gibrat's legacy', *Journal of Economic Literature*, **35**, 40–59.

APPENDIX 6.1: MULTILEVEL MODELS

The model applied in this chapter assumes that we have data from J groups, with a different number of respondents n_j in each group. On the respondent level, we have the outcome of respondent i in group j, variable Y_{ij}. There is an explanatory variable X_{ij} on the respondent level and one group-level explanatory variable Z_j. To model these data, a separate regression model in each group is formulated:

$$Y_{ij} = \beta_{0j} + \beta_{1j}X_{ij} + e_{ij} \tag{1}$$

The variation of the regression coefficients β_j is modeled by a group-level regression model:

$$\beta_{0j} = \gamma_{00} + \gamma_{01}Z_j + \mu_{0j} \tag{2}$$

and

$$\beta_{1j} = \gamma_{10} + \gamma_{11}Z_j + \mu_{1j} \tag{3}$$

The individual-level residuals e_{ij} are assumed to have a normal distribution with mean zero and variance σ_e^2. The group-level residuals $\mu_{0j} + \mu_{1j}$ are assumed to have a multivariate normal distribution with an expected value of zero, and they are assumed to be independent from the residual errors e_{ij}. The variance of the residual errors μ_{0j} is specified as σ_e^2 and the variance of the residual errors μ_{0j} and μ_{1j} is specified as σ_{u0}^2 and $\sigma_{\mu1}^2$. One can write this model as a single regression model by substituting Equations (2) and (3) into Equation (1). Substitution and rearranging terms gives:

$$Y_{ij} = \gamma_{00} + \gamma_{10}X_{ij} + \gamma_{01}Z_j + \gamma_{11}X_{ij}Z_j + \mu_{0j} + \mu_{1j}X_{ij} + e_{ij} \tag{4}$$

The segment $\gamma_{00} + \gamma_{10}X_{ij} + \gamma_{01}Z_j + \gamma_{11}X_{ij}Z_j$ in Equation (4) contains all the fixed coefficients; it is the fixed (or deterministic) part of the model. The segment $\mu_{0j} + \mu_{1j}X_{ij} + e_{ij}$ in Equation (4) contains all the random error terms; it is the random (or stochastic) part of the model. The term $X_{ij}Z_j$ is an interaction term that appears in the model because of modeling the varying regression slope β_{1j} of the respondent-level variable X_{ij} with the group level variable Z_j.

Even if the analysis includes only variables at the lowest (individual) level, standard multivariate models are not appropriate. Multilevel models are needed because grouped data violate the assumption of independence of all observations. The amount of dependence can be expressed as the intra-class correlation (ICC) ρ. In the random effect model the ICC is estimated by specifying an empty model, as follows:

$$Y_{ij} + \gamma_{00} + \mu_{0j} + e_{ij} \tag{5}$$

This model does not explain any variance in Y. It only decomposes the variance of Y into two independent components σ_e^2, which is the variance of the lowest-level errors e_{ij}, and $\sigma_{\mu 0}^2$, which is the variance of the highest-level errors μ_{0j}. Using this model, the (ICC) ρ is given by the equation:

$$\rho = \sigma_{\mu 0}^2 \,/\, (\sigma_{\mu 0}^2 + \sigma_e^2) \tag{6}$$

Our outcome variable Y_{ij} is the firm performance, measured as productivity. In the regression line (3) β_{0j} is the usual intercept, β_{1j} is the usual regression coefficient (slope) for the explanatory variable, and e_{ij} is the usual residual error term. The subscript j is for the region, and the subscript i is for individual firms. The difference with a usual regression model is that we assume that each region j has a different intercept coefficient β_{0j} and a different slope coefficient β_{1j} (since the intercept and slope vary across the regions they are often referred to as random coefficients).

7. Not being there: why local innovation is not (always) related to local factors

Richard Shearmur

INTRODUCTION

In this chapter I consider the links between territory and innovation, and draw together the implications of a number of recently published papers (Shearmur, 2010a, 2010b; Shearmur and Doloreux, 2009; Doloreux and Shearmur, 2011). Despite the popular notion that firm-level innovation is influenced by the *local* environment, I will argue that the scale at which the environment influences firm behavior is far from evident. Indeed, while there are many classic examples of successful local innovation systems (Baden Württemberg, Silicon Valley, Route 128, 3rd Italy, etc.), these may in fact be examples of well-located regions benefiting from the external linkages that their location provides. Although the end result is the same – firms in these regions are more successful and innovative than those in other regions – the causal mechanism suggested is one that is not connected with the locality itself. While this is beginning to be recognized by researchers (e.g. McCann, 2007; Andersson and Karlsson, 2004; Lorentzen, 2007), empirical evidence is still sparse. Implications for regional development policy have scarcely been explored.

In the particular cases mentioned above there probably *is* a local effect (due to the presence of urban agglomerations and of the myriad of opportunities and interactions they enable – Duranton and Puga, 2003), but this effect is combined with an accessibility effect (McCann, 2007). Since most studies of innovation systems focus on urban areas (Crevoisier and Camagni, 2001) and only look for local effects, the accessibility effect is usually not noticed. However, in non-metropolitan regions there may be no local effect: innovation in non-metropolitan regions may therefore vary according to the location of each region in relation to points of high accessibility (such as metropolitan areas (McCann, 2007)).

The chapter is structured as follows. The next section will briefly review the reasons why it is thought that there is a connection between firm-level innovation and the local environment. The third section will present some counter arguments and an alternative explanation for spatial variation in innovative activity, while the following section summarizes the overall argument and proposes a framework for understanding the connection between geographic space and innovation. The fifth section reviews some empirical results for Quebec that have been published elsewhere but whose wider implications have not been drawn out, and a paradox will be discussed: although, in general, innovation varies predictably with accessibility to metropolitan areas, this is not the case for high tech sectors (at least in the province of Quebec). These results have policy implications, particularly with regard to the indiscriminate application of development initiatives premised on the idea of localized innovation systems: the concluding section discusses these implications.

LOCAL INNOVATION SYSTEMS, LOCALITIES AND LOCAL DEVELOPMENT

Since the early 1980s, which were marked by a major recession in the Western world and by the recognition that developing countries were becoming increasingly competitive in a wide variety of industries, innovation has been seen as an important economic development strategy. Of course, there has always been innovation, and its effects have been particularly evident since the beginning of the industrial revolution. However, it is only with Solow (1956) that innovation (or technical change) was formally recognized as a key contributor to growth, and it is only as Western countries' favorable terms of trade with less developed countries eroded that policy makers focused upon innovation as the key to continued economic growth.

As innovation became a key dimension of macro-economic policy (see, for instance, Industry Canada, 2009), the ways in which innovation could be encouraged have increasingly come under scrutiny. Much of this work has focused upon economic sectors (Pavitt, 1984), upon establishment level behavior (Cohen and Levinthal, 1990) and upon innovation 'inputs' such as research and development and qualified workforce (Griliches, 1979; Baldwin and Johnson, 1995). The key unit of analysis was (and remains, in many cases) the establishment; and the micro-level motivation for innovation is understood either as cost-cutting or as maintaining (or enlarging) market share (von Hippel, 1988).

Increasingly, though, it has been noted that firms draw upon their environment to innovate (Cohen and Levinthal, 1990; Lundvall, 1992; Edquist, 1997): indeed, innovation is now often understood as a complex process with many feedback mechanisms, and firms interact with their surroundings in a variety of ways (Malecki, 1997). Thus, for an establishment to innovation, not only is it necessary for it to possess the appropriate absorptive capacities and internal resources, but it also needs to have access to an appropriate institutional and cultural framework that enables interactive learning, itself a key precondition of innovation (Lundvall, 1992). Much of this learning is based upon collaborative activity, exchanges of formal and informal information and appropriately flexible labor markets and legislative frameworks. This led Lundvall (1992) to put forward the idea of innovation systems, that is, each national context provides a particular mix of institutions, education, attitudes to novelty and so on, and some mixes – or systems – are more conducive to innovation than others.

In parallel to Lundvall's development of the idea of national systems of innovation, another body of work suggested the existence of *local* innovative milieu, somewhat along the lines of Marshall's industrial districts (Piore and Sabel, 1984; Maillat et al., 1993). Certain localities appeared to be better equipped than others to weather the industrial restructuring that occurred in the 1980s. This was attributed to the way in which small firms collaborated, shared a labor pool, exchanged information and reacted, as a system, to outside pressures. It is a short step from these examples to suggest the existence of local (or regional) innovation systems, which Cooke did in the mid-1990s (Cooke et al., 1997).

The 1990s saw a multiplication of concepts that posit a link between localities on the one hand and innovation and growth on the other. Whether clusters (Porter, 2003), learning regions (Florida, 1995), local innovative milieu (Maillat et al., 1993) or local innovation systems (Cooke et al., 1997) are considered, all suggest that regions possess certain tangible and intangible assets, and that these assets are drawn upon by local firms to generate innovation and growth. These ideas now form the mainstay of regional policy initiatives and are being widely disseminated (OECD, 2009).

The nature of these policies is somewhat uncertain. On the face of it they seem to be regional development policies: innovation, not necessarily world-firsts or patents but incremental product and process improvements, is thought to be necessary for a local economy to prosper and remain competitive. Thus innovation surveys are often conducted to provide information on the link between establishments and their environment (such as the Community Innovation Survey (OECD, 2006) and the Canadian Survey of Innovation (Statcan, 2006)). These surveys go beyond the particular type of innovation that is recorded in patent databases (Griliches, 1990) and record improvements to products and processes whatever their nature (OECD, 2005).

The surveys are designed to understand *how firms draw upon their environment in order to innovate*.[1] They are not designed to understand *the conditions under which innovation leads to regional development*. Causality in this direction is *assumed* at the regional level by (false) analogy with theories developed a-spatially by theoretical economists: indeed, while it is evident, following Solow (1956), that in a closed system innovation will lead to better factor productivity and therefore to growth, there is no guarantee that in an open system, such as a region, growth induced by innovation will be of local benefit.[2] Thus, while regional innovation policies may be appropriate if the objective is to increase overall levels of innovation, such policies should not be confused with regional development policies whose objective is to ensure that economic activity and employment remain in the region (Shearmur and Bonnet, 2010). In the rest of this chapter I will be investigating whether regions are an appropriate spatial unit over which innovation policies should be deployed. The question of innovation's *impact* on regions will not be addressed. The types of innovation that will be referred to are product and process innovations – often, but not necessarily, incremental in nature – that are measured in innovation surveys.[3]

BORROWED MILIEU? INNOVATION AND NON-LOCAL EXTERNAL FACTORS

One of the problems with the idea that certain regions can enhance local innovation by virtue of their local characteristics (whether clusters, institutions, labor markets, etc.) is that it is partial and incomplete: and while a partial and incomplete idea is fine in an academic setting because it invites critical review and discussion, which slowly and elliptically push forward the boundaries of our knowledge (and, of course, reveal our ignorance), it becomes problematic when it is adopted by policy makers. The problem, in a policy context, is twofold. First, public money may be spent on ineffective interventions and, second, the promises being made may not be kept.

One of the key problems with the idea that local factors are directly connected with local innovation has been raised in the wider literature on regional development. Indeed, in this context it has been shown that the connection between local attributes and local growth is often rather weak: endogenous growth theory, theorized and tested at the international scale (Lucas, 1988; Romer, 1989), and, to some extent, at the inter-metropolitan scale (Simon, 1998; Glaeser et al., 1995), does not necessarily apply to small open regional economies. Not only do such economies not necessarily capture the effects of their own factor endowments, they are also subject to external structural forces that overwhelm local effects.

Indeed, it can be argued that over most periods and in most contexts geostructural factors such as density, accessibility to major markets and industrial structure have a strong effect on local growth trajectories for most regions (Shearmur and Polèse, 2007; Shearmur et al., 2007; Davis and Weinstein, 2002), with local measurable attributes (such as costs and human capital) having a marginal effect. The whole field of new economic geography (Krugman, 1995;

Fujita and Mori, 2005) is based upon the idea that regions grow and decline in response to factors such as agglomeration, transport costs and technological changes that tend to operate irrespective of local particularities. There is no reason to believe that the innovative capacity of regions is not affected by the same geostructural factors that influence their economies as a whole.

Although the empirical results that underscore the role of geostructural factors leave plenty of room for local non-measured effects (such as institutions, culture and human capital), they also show that geostructural factors cannot be ignored: the location of a region relative to other regions is a key determinant of employment, population and income growth (and, I suggest, of innovation dynamics) irrespective of other factors. This view is not shared by all economic geographers: indeed, Storper and Walker (1989), for instance, argue strongly that it is local attributes that have the predominant effect. However, this influential view is partly premised on comparisons between nations, and partly on a discussion of the fate of certain major cities: and while it may be correct at such scales where spatial units are (relatively) closed systems, it does not apply to sub-national or non-metropolitan regions where, at the very least, it is a combination of local and non-local factors that must be considered.

Let us consider the difference between the development trajectories of two small towns. Although *local* attributes such as human capital or collaborative behavior may be relevant, *accessibility to non-local* markets and to important actors and institutions are of crucial importance. A small town well located relative to a metropolitan area may 'borrow the system'[4] of its larger neighbor, leading to local firms innovating not by virtue of any local opportunities or dynamics but rather because of the town's accessibility to a location that is a strong source of external factors of innovation. Conversely, an identical small town located in remote northern Canada (for instance) would probably experience great difficulty in generating sustained innovative behavior because of the lack of access to key contacts and learning opportunities that proximity to a large city provides: the remote town has no nearby system to 'borrow'.

Unfortunately, the Storper and Walker (1989) approach, while valuable as one possible way of understanding differences in local economic growth, has tended to dominate debates over the last two decades in the geographic disciplines (Asheim and Gertler, 2006). In addition there has been an almost wholesale rejection of quantitative techniques and of spatial analysis over this period (Tickell et al., 2007; Barnes, 2009). Thus, during the very years when the connection between innovation and geography was being analysed, the conceptual framework and the analytical methods in economic geography have been geared towards showing that local dynamics can explain local growth and innovation processes. Wider dynamics, those that function at a scale beyond the region and are therefore difficult to identify through locally based case studies, have often been overlooked by economic geographers. In particular, the idea that firms in certain localities and regions may 'borrow the milieu' from other places has not really been considered.

Recently, a number of important ideas have been suggested that add meat to the bones of this critique. They will not all be covered here, but four important contributions will be outlined. Bathelt et al. (2004) suggest that innovation systems (i.e. urban areas that benefit from local buzz) are not free-standing. They are connected to other systems through 'pipelines' – privileged channels of communication, transport and often of social proximity – that enable synergies to occur between different innovations systems. This important contribution basically applies Castells' (1996) conceptualization of the globalized world to innovative milieu: this is a world consisting of a series of urban nodes connected to each other by a space of flows – both virtual (electronic, internet, etc.) and physical (road, air or high-speed rail connections). These physical connections remain important because even with virtual connectivity the exchange of tacit knowledge and the building of trust still require face-to-face contact. This idea moves beyond a simple connection between local attributes and local innovation, since firms in one milieu can interact easily, via pipelines, with those in others. However, this conceptualization still does not take into account the spaces – and there are many of them – which lie outside of buzzing milieu, spaces which include smaller cities and rural or peripheral regions: innovation is still understood as occurring primarily in large cities because these areas provide the necessary local conditions and provide good access to other similar regions through the pipelines.

While Bathelt et al. (2004) take one step away from the idea that physical proximity between actors is a necessary condition for knowledge exchange and collaboration, Boschma (2005) goes somewhat further in conceptualizing the different types of proximity that may lead to innovation. He identifies five types of proximity: (i) cognitive (sharing a common vocabulary and conceptual framework); (ii) organizational (capacity to coordinate and exchange knowledge); (iii) social (micro-level social ties of friendliness and trust); (iv) institutional proximity (macro-level routines, rules and regulations); and (v) geographic proximity. Thus, geographic proximity between actors is seen here as only one of a variety of ways that information exchange and innovation may be stimulated. Indeed, it is possible to push this idea further. It could be argued that the only reason geographic proximity plays a role in stimulating innovative behavior is that it facilitates the four other types of proximity. If it is possible to conceive of a way in which the other types of proximity can be maintained across space, then the idea that the external factors conducive to innovation are necessarily local will need to be questioned.

Torre (2008) suggests a way in which the first four types of proximity can be maintained without the need for co-location. He recognizes that the types of proximity identified by Boschma (2005) are necessary preconditions for successful interaction, knowledge exchange and collaboration between actors. Furthermore, another type of interaction – fortuitous or chance encounters – can also be important. Torre (2008) argues that, at least for some establishments, these proximities and fortuitous encounters can be adequately created and maintained through temporary proximity. It may suffice that actors be brought together temporarily (in, say, fairs, conferences, meetings or temporary collaborations), which may then be further built upon by visits and staff exchange, for successful innovative dynamics to be created. Often conventions and fairs will gather, in one location, actors who share at least one of Boschma's (2005) four non-geographic proximities, and temporary geographic co-location enables the informal exchanges and chance encounters between like-minded actors that provide the external stimulus to innovation.

Finally, McCann (2007) has built a theoretical model which accounts for the distribution across space – at an inter-regional scale – of different types of innovation. He starts with the idea that the co-location of actors is not necessary for all types of innovation. He proposes that different types of innovation have different requirements in terms of intensity and frequency of interaction. In this he is in line with Torre's (2008) idea that, at least for some types of innovation, temporary proximity suffices. A further hypothesis of the model is that, all else being equal, large buzzing metropolitan areas are the locations in which the most frequent and intense interactions can take place. Given these starting points he shows that the types of innovation requiring the most frequent and intense interactions will tend to occur in metropolitan areas. Through a process of crowding out, innovations that require lesser intensities or frequencies of interaction will take place further away from metropolitan areas. There will thus be a concentric distribution of innovative activity around metropolitan centers, with the most radical and interaction-intensive innovations requiring a central location, and the least interaction-intensive innovations occurring furthest away. Duranton and Puga (2003), come to a compatible conclusion using a different type of reasoning (based upon product life cycles). They suggest that the most radical innovations (the development of new products for new markets) will occur in large buzzing metropolitan areas, which they call nursery cities, whereas less radical cost-saving ones (such as process innovation) will tend to occur in smaller specialized cities: although they still see innovation as a purely urban phenomenon, they differentiate between different sorts of innovation and suggest that these will follow a spatial logic not necessarily connected with local innovation systems.

If the various ideas just outlined are correct, then there is no reason to suppose that an establishment's *local* environment is the only (or even principal) external determinant of its innovative behavior, although in some cases it may be. The geographic co-location between actors, institutions and potential collaborators is not a *necessary* environmental factor that determines innovative behavior. Innovation systems can operate across space and are not necessarily bounded in any simple way by proximity.[5] Geography still matters, because clearly accessibility to metropolitan areas is important: but 'being there' (Gertler, 2003) is not.

In a recent paper, Lundvall (2005, p. 7) makes a related point. He states that:

> the direct application of the core element of the innovation system – interactive learning – to the regional level [is] problematic. Much of the relevant interaction takes place at the national/international rather than at the regional level and other 'systemic' mechanisms may be more important when it comes to explain the formation and evolution of regional clusters.

The questions of scale and regional size, which underpin Lundvall's statement, have not been adequately resolved in the regional innovation literature. If the elements of an innovation system can only be brought together at a national scale, or only in sufficiently large regions, what is the geographic logic of innovation within nations or in spaces between large regions? It is to this question that the chapter now turns.

SUMMARY: A GEOGRAPHY OF INNOVATION

Bearing in mind the literature just reviewed, the geographic pattern of innovation that is suggested in this chapter can be understood by referring to Figure 7.1. With this schematic geographic system in mind, the spatial logic of innovation is as follows.

While of course dependent upon their internal absorptive capacities, their investment in research and development and their dynamism, the innovative activity of establishments is also dependent upon their location. It will vary along two independent geographic dimensions.

1. It will vary between *localities* according to the *nature and type of local buzz*. This is the classic local innovation systems approach and is closely connected with the idea of endogenous development. However, not all areas possess buzz. In Figure 7.1, only the large cities generate buzz, and maybe – but not necessarily – the smaller cities.
2. It will vary between *locations* according to each location's *accessibility to buzz*. This does not mean that innovation will decrease with distance from large cities. As McCann (2007) points out, different types of innovation require different levels and intensities of interaction. Thus, one would expect the *type of innovation* and the *process of innovation* to differ with distance from large cities (and maybe with distance from small ones). For instance, if Duranton and Puga's (2003) theory is correct, non-radical

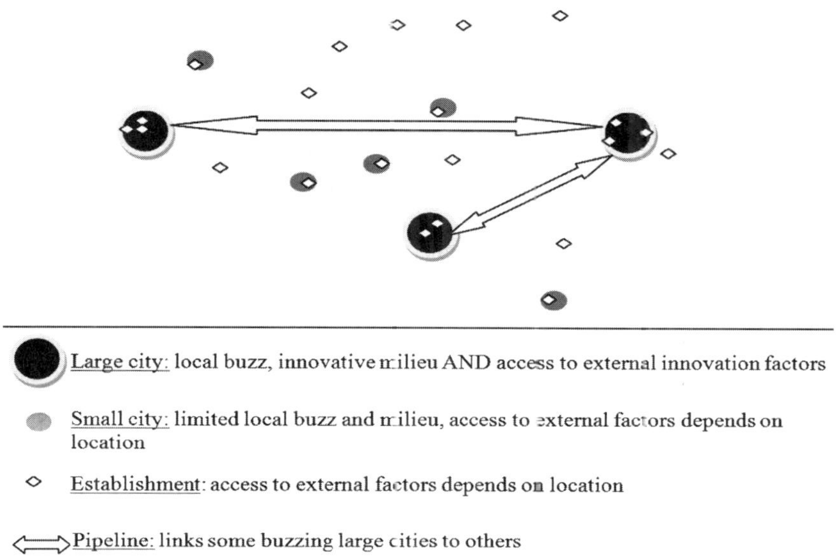

Figure 7.1 A framework for understanding the geography of innovation

process innovation will occur further away from large cities than radical product innovations, and will maybe occur in small urban areas.

In this view of the world, distance and transport networks become important considerations for understanding regional innovation activity. Figure 7.2 provides a summary of the different types of regional environment that can be envisaged.

1. *High interaction environments* are those where it is most likely that innovations will occur. This is because innovation depends not only on the local innovation system (which is well developed in these environments) but also on accessibility to outside factors (which is also well developed in these environments). Innovation dynamics in these environments have tended to be attributed exclusively to the local innovation system (Asheim and Gertler, 2006; Storper and Walker, 1989): however, Figure 7.2 illustrates that it is both local buzz *and* accessibility that lead to their high interaction levels (and hence innovativeness) (Bathelt et al., 2004).
2. *Medium interaction environments* are those in which only one of the two enablers of interaction, local buzz or accessibility, is present.

- Some medium sized cities of appropriate size and with appropriate local institutions may manage to generate local buzz (upper left quadrant of Figure 7.2). However, this local buzz is not complemented by easy interaction with outside actors. The isolation of these medium cities makes integration into wider innovation processes difficult: the innovation performance of these local systems may appear deficient, but it has nothing to do with local culture, institutions or other such factors. Rather, improved accessibility may be required in order to induce synergies and complementarities.
- Some small towns or rural areas may generate little or no local buzz, but may be close (or well connected) to major metropolitan areas (lower right quadrant of Figure 7.2). The absence of local factors of innovation can be partly overcome by drawing upon outside factors since accessibility to major metropolitan areas is not problematic: because no local buzz is generated, these regions will appear

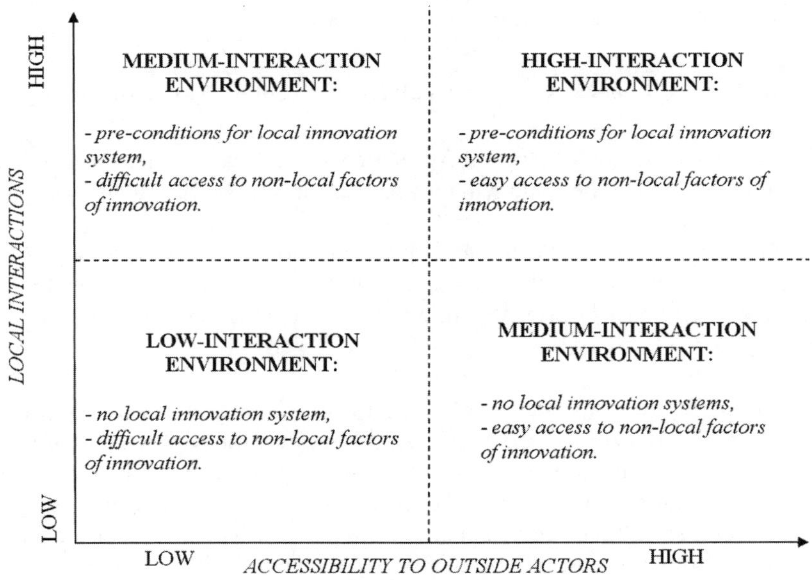

Note: This typology classifies the location of establishments not of regions: each point in geographic space can be positioned somewhere along both axes, and each point within a region (however defined) will have a somewhat different position in this two-dimensional system

Figure 7.2 A two-dimensional typology of establishment location: local interactions and accessibility to non-local actors

to overperform if local levels of innovation are thought to emanate purely from local factors and institutions.

3. Finally, *low interaction environments* are those places that benefit from neither local buzz nor accessibility: they are too small or low density to generate local innovation processes, and too isolated for establishments there to enter into close interaction and collaboration with actors from the outside. From neither the innovation systems nor the accessibility perspective are these environments expected to be conducive to innovation. If current regional innovation policies are followed, the solution to overcoming the poor innovation capacity of such localities would be to strengthen local institutions and knowledge infrastructure, even if the preconditions for implementing such an approach are virtually absent (Doloreux and Shearmur, 2006). From the perspective suggested in this chapter, an alternative approach for enabling innovation in this type of region would be to improve accessibility to regions where institutions, knowledge and interactions are available. This means good transport and good communications technology, but also transregional institutions and events capable of ensuring the introduction of entrepreneurs to appropriate contacts in distant milieu. These institutions and events would aim at generating social or organizational proximity (Boschma, 2005) by way of temporary proximity (Torre, 2008).

IS THERE ANY EMPIRICAL EVIDENCE OF A CONNECTION BETWEEN ACCESSIBILITY AND INNOVATION?

In a recent series of papers (Shearmur and Doloreux, 2009; Shearmur, 2010a, 2010b) we have put some of these ideas to the test. The purpose of this section is to briefly present the model and highlight a few key findings: full details of the empirical results can be found in the cited papers.

Using data for the province of Quebec[6] (see Figure 7.3), the spatial distribution of different types of innovation relative to Quebec's major metropolitan areas has been studied, both for knowledge-intensive business services (KIBS) (Shearmur and Doloreux, 2009) and manufacturing establishments (Shearmur, 2010a, 2010b). The KIBS data are a 20% sample of all KIBS establishments in Quebec (1124 observations), and the manufacturing data come from a census of all establishments of over 20 employees (3200 observations).

The province of Quebec is the size of a small European country. It comprises 7.8 million people, and, despite the large distances evident from Figure 7.3, most of these people reside between Quebec City and Ottawa-Gatineau. Without being equivalent to a high density European country, the majority

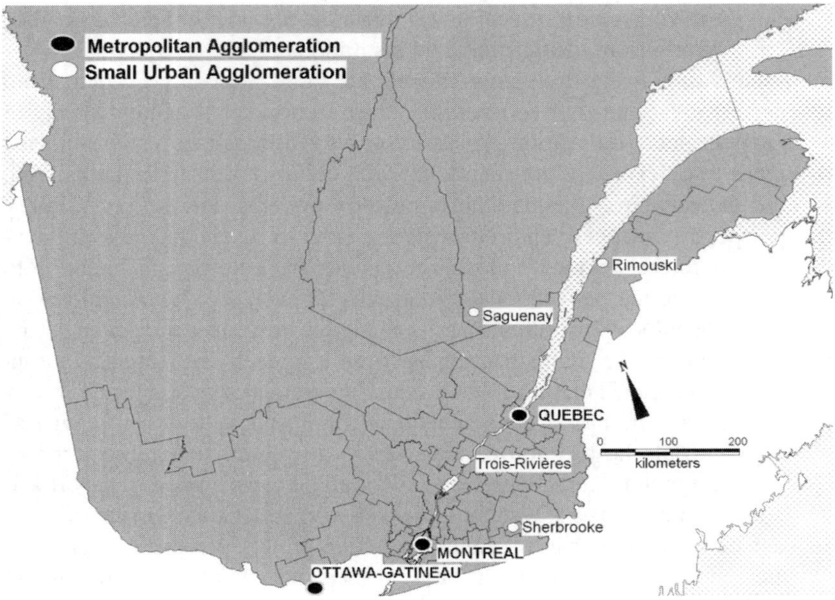

Figure 7.3 Quebec, Canada: simplified urban system and regional boundaries

of Quebec's economic activity takes place in a relatively compact area and is fairly similar to a country like Sweden. The province of Quebec has considerable autonomy from Canada: it raises its own income tax (which corresponds to about 50 percent of all income tax paid by Quebecers), manages its own education system and sets its own industrial and development policies. It also has its own language, French, in which all day-to-day business is conducted. In most respects (except for foreign policy and for equalization paid out by the federal government) Quebec has powers that resemble those of a nation state. To test the model, Quebec's three largest cities are classified as large metropolitan areas, and four smaller cities are classified as small urban areas.

A simple logistic regression is applied to a binary variable (innovates/does not innovate). The innovation questions follow the *Oslo Manual* (OECD, 2005) model, but with managerial and marketing innovations also included in the KIBS questionnaire. After a series of control variables are placed in the model, the distance to large and small cities is entered, and its effect upon the probability of innovating is observed.

Formally, the model is shown in Equation (7.1) as follows:

$$I_i^x = f(C_i, D_{iy}, R_i) \qquad (7.1)$$

where I_i^x is a binary variable indicating whether or not establishment i has introduced innovation type x. For manufacturing establishments, four types of innovation are analysed: two product innovations (new only to the establishment; new to Quebec or beyond) and two process innovations (new only to the establishment; new to Quebec or beyond). For KIBS establishments, an aggregate measure of minor and major innovations is introduced. C_i is a vector of control variables including establishment size, exports and three digit NAICS sector. These control variables are connected with establishment level innovation and are also known to vary across space (Polèse and Shearmur, 2002): they therefore eliminate 'obvious' spatial factors, such as the effect of industrial structure on spatial patterns of innovation. D_{iy} is a vector of distance variables. For each establishment i, the following distances are included: d_{im} is the distance to the closest metropolitan area of over 500 000 people (Québec City, Montreal, or Ottawa); $d_{im}{}^2$ is the square of d_{im}, d_{is} is the distance to the closest small urban areas (i.e. an urban area of over 50 000 people located further than 100 km from a major metropolitan area – this comprises Sherbrooke, Saguenay, Trois-Rivières and Rimouski. Each of these cities has its own university); $d_{is}{}^2$ is the square of d_{is}; R_i is the regional fixed effects. Québec is subdivided into 50 regions based upon MRC (county) boundaries. Some have been aggregated in order to ensure at least 20 observations by unit. It should be noted that except in immediate proximity to Montreal, labor markets do not usually overlap with MRC boundaries. MRCs are administrative units and correspond to the scale at which many regional development and employment policies are delivered.

A DISCUSSION OF RESULTS

Whether KIBS or manufacturing establishments are considered, one key result emerges from these analyses: regional fixed effects provide little explanation for the spatial variation in establishments' propensity to innovate. Distance from major metropolitan areas, on the other hand, does have some explanatory value. Thus, the idea that establishment-level innovation varies across Quebec according to local institutional, cultural and other conditions is not supported, *unless* these also happen to vary in a continuous fashion with distance from Quebec's larger cities. McCann's (2007) model, on the other hand, is corroborated: as accessibility to metropolitan areas varies so does the intensity of different types of innovative activity.

Another key result emerges from the manufacturing data (Shearmur, 2010a). Each type of innovation reaches a maximum probability at a specific distance from large and small cities (multiple maxima are possible because of the quadratic nature of the models and because both large and small cities are considered). Table 7.1 shows the distances at which each type of innovation reaches its maximum probability.

Table 7.1 Location of peak probabilities of establishment-level innovation, Quebec 2005

	Location of innovation maxima relative to closest metropolitan area	Location of innovation maxima relative to closest small city
Radical product	0 km	no effect
Product new to firm	15 km	no effect
Radical process	30 km	no effect
Process new to firm	35 km	0 km

Note: Full results available in Shearmur (2010a)

According to McCann's (2007) model, the observed maxima (Table 7.1) suggest that radical product innovations are the most interaction-intensive, followed by product innovations new to the firm, then radical process innovations and, finally, process innovations new to the firm. This classification of innovation types according to their need for interactions is plausible, and the results corroborate the idea that establishments which exhibit innovation activities requiring less frequent and intense interactions will locate further away from metropolitan areas.

Only minor process innovations exhibit any association with small urban areas: in keeping with Duranton and Puga's (2003) idea (based upon product lifecycles and cost-cutting), these minor cost-cutting innovations show a higher probability of occurrence in Quebec's smaller cities.

These results are derived from the study of all manufacturing establishments. Given that the McCann (2007) model is loosely based upon the idea that manufacturers compete for the best locations, it is not to be expected that they will be easily replicated when observing specific manufacturing sectors: indeed, although it is reasonable to suppose that competition for locations (and access to interactions) is confined to the universe of manufacturing establishments (i.e. they compete for real-estate between themselves but not with establishments in other sectors), there is no reason to believe that it is confined to establishments within specific manufacturing sectors. Having said this, the results for first and second transformation and for medium tech manufacturing confirm the pattern just described (Shearmur, 2010b). However, a curious result is found for high-tech manufacturing, one which is also observed for certain high-order services: the probability of innovating in these establishments actually *rises* with distance from metropolitan areas.

Why would innovation in establishments in high-tech sectors increase with distance from large buzzing cities? Three possible interconnected explanations are forthcoming (Shearmur, 2010b; Doloreux and Shearmur, 2011):

1. The highest densities of high-tech establishments are found in and around large metropolitan areas (unlike for lower-tech establishments which are more evenly spread out across space). The probability of innovating rises with distance from a large city (for high-tech establishments) because only those establishments that innovate can survive at a distance from high-interaction locations. This is not the case for lower-tech establishments which have good cost- and resource-based reasons to locate at a distance from metropolitan areas.

   From this perspective there are two different spatial logics to innovation. In sectors where firms derive a cost advantage from locating away from metropolitan areas, then only innovators will be prepared to pay the higher costs associated with metropolitan proximity because this proximity is necessary for their innovative activities. In high-value-added sectors where the cost advantage of remoter locations is virtually non-existent, there is no economically rational reason to locate away from a metropolitan area. The *existence* of high-tech establishments in these regions therefore depends principally upon non-economic factors (such as lifestyle choice or the entrepreneur's place of residence). However, because they must overcome the disadvantage of remoteness, the *survival* of such firms rests upon their capacity to innovate, leading to the higher propensity to innovate that we observe.
2. In high-tech sectors not all firms innovate. In metropolitan areas large integrating firms in high-technology sectors may produce major innovations by combining and re-combining standard products made by other firms in similar or related sectors. The suppliers of these standardized inputs will not necessarily innovate *internally*, but will participate as collaborators or suppliers to major innovative projects conducted by the integrator. In Montreal, for instance, the aircraft manufacturer, Bombardier, is an example of such an integrator: this large firm produces multiple innovations, while its suppliers may not produce many. If such networked innovation is more frequent in large metropolitan areas, the probability that any specific establishment will innovate there will be lower (though of course many innovations, such as all those produced by Bombardier, occur in the city).

   If this interpretation is correct, then it reinforces the idea that there exist collaborative and/or supply chain connections within metropolitan areas, particularly in high-tech sectors. Innovation is the output of a network of firms, with a low number of firms declaring an innovation (and maybe one or two major firms introducing multiple innovations). Conversely, as one moves away from metropolitan areas, high-tech firms internalize their innovative activity (maybe because of the lack of potential collaborators and suppliers): this internalization means that more firms will declare that they have introduced an innovation.

3. Finally, the higher probability of innovation for high-tech establishments in remoter locations may be due to choice: some cutting-edge firms may have no desire to share information and knowledge with others. Secrecy – facilitated by relative remoteness from buzz and milieu – may be sought by some cutting-edge innovators, and this may also explain the higher probability of finding innovative high-tech firms in remoter locations.

Before moving on to some concluding remarks, it is important to emphasize that although the geographic patterns discussed above are statistically significant over and above the control variables, the inclusion of distance variables usually does not increase the explanatory power of the models very much. In Quebec, while regions (counties) have virtually no independent explanatory power when it comes to understanding the geography of innovation, location relative to large cities has significant (but not particularly strong) explanatory power. If one wishes to understand in absolute terms why establishments in one region (or location) are more or less innovative than in another, then industry is a key factor: those regions which house establishments in the most innovative sectors produce the most innovations. Geography may matter, but, for innovation, only at the margin.

POLICY IMPLICATIONS

In this chapter the connection between geography and innovation as defined and measured in the *Oslo Manual* (OECD, 2005) has been discussed. Broadly speaking, two different (and possibly complementary) mechanisms have been described: on the one hand, the well-rehearsed idea that innovation activity draws upon, and is enabled by, the local environment (institutions, local suppliers and clients, local laborforce, and so on); on the other hand, the less well-known proposition that innovation activity is not connected to the local environment but depends upon access to locations which are high-points of interaction and entry points to global networks.

At one level it could be argued that these differences are almost trivial: unless one is a geographer (or a regional development officer), whether or not a firm draws upon its *local* environment or upon an environment which it has *access to* is of little importance. The real questions for researchers interested in establishment-level innovation are *whether* a firm relies upon external factors and, if so, *in what way*.

From a regional or local development policy perspective, however, the geography of these external factors is crucial. For a variety of reasons associated with the history of ideas in economic geography, particularly with the rejection of quantitative spatial analysis and the concomitant argument that each region and locality develops its own institutions and particular brand of capitalism (Storper and Walker, 1989; Gibson-Graham, 1996), the connection between innovation and space has been understood almost exclusively as a story of local embeddedness and of the evolution of local innovation systems. This approach is also in keeping with ideas on local development, themselves strongly supported by the (mis)application of endogenous growth theory to all types of regions, even those too small to generate and capture endogenous processes.

If the analysis of innovation and space is limited to metropolitan areas and large cities, this local approach seems to apply. There is indeed some connection between the endowments and institutions found in metropolitan areas and their innovative behavior (Simmie, 2001; Acs, 2002), although the precise nature of this connection remains somewhat elusive (Hall, 1999). The 'localities' approach works because metropolitan areas are large enough to develop and sustain at least some of the institutions and interactions necessary for interactive learning to occur (Lundvall, 2005).

There is no reason, however, to suppose that the local approach will have much to say about establishment-level innovation in smaller cities and rural areas where such institutions and interactions cannot be sustained.

The discussion in this chapter, and the results that have been described, point to another process that may explain the connection between innovation and space, not only in metropolitan areas (where it is combined, and often confused, with endogenous effects), but also in non-metropolitan areas where the low density and low diversity of actors make the localities approach more problematic. In these areas, *local* endowments and institutions have less impact upon establishment-level behavior than *accessibility* to other regions where these endowments and institutions can be found. This spatial pattern of innovation has not been noticed, I suggest, because few people have looked for it: the dominant approach to understanding innovation and space has been by way of local milieu and local systems, and spatial patterns and theories have only recently been explored at the margins of the field.

From a regional policy perspective – assuming that establishment-level innovation is sought – these results have important implications. The approach which currently dominates is to strengthen local institutions and build up territorial capacity (OECD, 2009): in some regions, particularly large or central cities,[7] this may be a good approach. Within such cities local capacity building may enable local establishments to interact among themselves, leading to collaborations and interactive learning. Further away from large or central cities, local capacity building may be less effective: the density of local actors is often simply too low, and a precondition of interactive learning – wide and varied networks – is absent. In these regions an effective innovation policy may be to improve connections (both physical, by way of transport, and social, by way of conferences, fairs and other focused networking opportunities) with outside actors with whom synergies may develop. In other words, for some regions the resources for innovation will come almost entirely from the outside, and a successful policy approach would facilitate access to them rather than try to develop them locally.

Finally, and still from a regional perspective, a note of caution should be sounded about innovation policies. At a wide scale – that is, at the scale of an economy that can internalize many or most economic processes – innovation is quite correctly perceived as an important factor of growth. National level governments support and encourage innovation on the basis that it is necessary for 'local' (i.e. national) firms to innovate in order for them to remain competitive and to open up new markets. If jobs are destroyed in one industry, they will usually be created in another, in a Schumpeterian process of creative destruction. To some extent this also occurs within the economies of large metropolitan areas.

In smaller cities and regions, usually more specialized, such virtuous processes do not necessarily occur (Shearmur and Bonnet, 2010). Innovation is no doubt necessary for the survival of local establishments, but innovation can lead to job losses, either through productivity gains or through establishment relocation as access to factors of production, services and wider networks become more important. At a fine regional scale *there is no guarantee that new jobs will be created locally to replace the old*. New establishments will certainly emerge somewhere in the economy, but there is no reason to believe that they will emerge in the same locality where innovation has led to job losses.

This is another way of pointing out the limits of viewing small regions and cities through the lens of endogenous growth theory: this theory, successfully applied to explain growth differences between nations, may also explain growth differences between large urban areas. However, the smaller the region (in terms of number and density of actors), the less it can internalize economic processes and the more endogenous growth theory – or theories associating territory and innovation – breaks down. Lundvall (2005), who suggests that the innovation system approach may break down if it is applied at the sub-national level, expresses a very similar point.

NOTES

1 The surveys also examined the use of resources internal to the firm for innovation purposes, but this dimension is not relevant to the discussion here.
2 An example will illustrate this point: in Canada there are many single industry towns. When the local industry innovates (often by introducing productivity-enhancing process innovations) there are fewer jobs. Since there are no other industries to employ the out-of-work labor, the localities themselves decline even as the industries in them become more productive. In this example local innovation (which increases overall utility) has dire consequences for the locality. The connection between innovation and local development is thus not a necessary one, as is often assumed, and calls for further detailed study.
3 I will not be discussing patents or the body of work that examines the spatial diffusion of citations (e.g. Jaffe, 1989; Breschi, 1999; Breschi and Lissoni, 2004).
4 Alonso (1973) uses the term 'borrowed size', to describe how small cities that are accessible to larger ones may benefit from the agglomeration economies of their more or less distant neighbors. This idea is discussed by Phelps et al (2001) in the context of small firms and their shift to rural areas. I am suggesting here that the concept may also apply to milieu and local innovation system effects.
5 A growing number of papers have been providing empirical support for this contention but they have not yet been considered as a coherent body of work (see, for instance, Echeverri-Carroll and Brennan (1999), Britton (2003) and Lorentzen (2007)). This chapter is an attempt to construct a framework that can encompass these results.
6 The data are fully described in the cited papers.
7 Central cities have high accessibility to a large hinterland and are a focal point of this hinterland's activity: this may provide sufficient opportunities for interaction and interactive learning as actors temporarily co-locate there to conduct business.

REFERENCES

Acs, Z. (2002), *Innovation and the Growth in Cities*, Cheltenham, UK and Northampton, MA, USA: Edward Elgar Publishing.

Alonso, W. (1973), 'Urban zero population growth', *Daedalus*, **102**, 191–206.

Andersson, M. and C. Karlsson (2004), 'The role of accessibility for the performance of regional innovation systems', in C. Karlsson, P. Flensburg and S.-A. Horte (eds), *Knowledge Spillovers and Knowledge Management*, Cheltenham, UK and Northampton, MA, USA: Edward Elgar Publishing, pp. 283–310.

Asheim, B. and M. Gertler (2006), 'The geography of innovation: Regional innovation systems', in J. Fagerberg, D. Mowert and R. Nelson (eds), *The Oxford Handbook of Innovation*, Oxford: Oxford University Press, pp. 291–317.

Baldwin, J. and J. Johnson (1995), *Human Capital Development and Innovation: The Case of Training in Small and Medium-Sized Firms*, Ottawa: Statistics Canada, working paper 11F0019MPE No. 74.

Barnes, T. (2009), '"Not only ... but also": Quantitative and critical geography', *The Professional Geographer*, **61** (3), 292–300.

Bathelt, H., A. Malmberg and P. Maskell (2004), 'Clusters and knowledge: Local buzz, global pipelines and the process of knowledge creation', *Progress in Human Geography*, **28** (1), 31–56.

Boschma, R. (2005), 'Proximity and innovation: a critical assessment', *Regional Studies*, **39** (1), 61–74.

Breschi, S. (1999), 'Spatial patterns of innovation: evidence from patent data', in A. Gambardella and F. Malerba (eds), *The Organization of Innovation in Europe*, Cambridge: Cambridge University Press, pp. 71–102.

Breschi, S. and D. Lissoni (2004), 'Knowledge networks from patent data', in H. Moed, W. Glanzel and U. Schmoch (eds), *Handbook of Quantitative Science and Technology Research*, Amsterdam: Springer, pp. 613–43.

Britton, J. (2003), 'Network structure of an industrial cluster: electronics in Toronto', *Environment and Planning A*, **35** (6), 983–1006.

Castells, M. (1996), *The Rise of the Network Society*, Oxford: Blackwell.

Cohen, E. and D. Levinthal (1990), 'Absorptive capacity: a new perspective on learning and innovation', *Administrative Science Quarterly*, **35**, 128–52.

Cooke, P., M. Gomez Uranga and G. Etxebarria (1997), 'Regional innovation systems: institutional and organizational dimensions', *Research Policy*, **26** (4–5), 475–91.

Crevoisier, O. and R. Camagni (2001), *Les Milieu Urbains: Innovation, Systèmes de Production et Ancrage*, Neuchâtel: EDES.

Davis, D. and D. Weinstein (2002), 'Bones, bombs, and break points: the geography of economic activity', *The American Economic Review*, **92** (5): 1269–89.

Doloreux, D. and R. Shearmur (2006), 'Regional development in sparsely populated areas: the case of Quebec's missing maritime cluster', *Canadian Journal of Regional Science*, **29** (2), 195–220.

Doloreux, D. and R. Shearmur (2011), 'Collaboration, information and the geography of innovation in knowledge intensive business services', *Journal of Economic Geography*, doi: 10.1093/jeg/lbr003.

Duranton, G. and D. Puga (2003), 'Nursery cities: urban diversity, process innovation, and the life-cycle of products', *The American Economic Review*, **91**(5), 1454–77.

Echeverri-Carroll, E. and W. Brennan (1999), 'Are innovation networks bounded by proximity?', in M. Fischer, L. Suarez-Villa and M. Steiner (eds), *Innovation, Networks and Localities*, Berlin: Springer-Verlag, pp. 28–49.

Edquist, C. (1997), *Systems of Innovation: Technologies, Institutions and Organizations*, London: Pinter.

Florida, R. (1995), 'Toward the learning region', *Futures*, **27**, 527–36.

Fujita, M. and T. Mori (2005), *Frontiers of the New Economic Geography*, discussion paper no. 27, Kyoto: Institute of Developing Economies.

Gertler, M. (2003), 'Tacit knowledge and the economic geography of context, or The undefinable tacitness of being (there)', *Journal of Economic Geography*, **3**, 75–99.

Gibson-Graham, J-K. (1996), *The End of Capitalism (As We Know It)*, Oxford: Blackwell.

Glaeser, E., J. Scheinkman and A. Shleifer (1995), 'Economic growth in a cross section of cities', *Journal of Monetary Economics*, **36**, 117–43.

Griliches, Z. (1979), 'Issues in assessing the contribution of research and development to productivity growth', *The Bell Journal of Economics*, **10** (1), 92–116.

Griliches, Z. (1990), 'Patent statistics as economic indicators: a survey', *Journal of Economic Literature*, **92**, 630–53.

Hall, P. (1999), *Cities in Civilization*, London: Phoenix Books.

Industry Canada (2009), *Business Plan 2009–2010*, Ottawa: Industry Canada, http://www.ic.gc.ca/eic/site/ic1.nsf/vwapj/IC_BusinessPlan200910-eng.pdf/$file/IC_Business Plan200910-eng.pdf (accessed 18 September 2009).

Jaffe, A. (1989), 'Real effects of academic research', *American Economic Review*, **79**, 957–70.

Krugman, P. (1995), *Development, Geography and Economic Theory*, Cambridge, MA: MIT Press.

Lorentzen, A. (2007), 'The geography of knowledge sourcing – a case study of Polish manufacturing enterprises', *European Planning Studies*, **15** (4), 467–86.

Lucas, R. (1988), 'On the mechanics of economic development', *Journal of Monetary Economics*, **22**, 3–42.

Lundvall, B-Å. (1992), *National Systems of Innovation: Towards a Theory of Innovation and Interactive Learning*, London: Frances Pinter.

Lundvall, B-Å. (2005), *National Innovation Systems – Analytical Concept and Development Tool*, 10th DRUID conference, Copenhagen, http://www.druid.dk/conferences/summer2005/papers/ds2005-603.pdf.

Maillat D., M. Quévit and L. Senn (1993) (eds), *Réseaux d'innovation et milieu innovateurs: un pari pour le développement régional*, GREMI/EDES, Neuchâtel.

Malecki, E. (1997), *Technology and Economic Development: The Dynamics of Local, Regional and National Competitiveness*, London: Addison Wesley Longman.

McCann, P. (2007), 'Sketching out a model of innovation, face-to-face interaction and economic geography', *Spatial Economic Analysis*, **2** (2), 117–34.

OECD (2005), *Oslo Manual* (3rd edn), Paris: Organisation for Economic Co-operation and Development and Eurostat.

OECD (2006), *Community Innovation Statistics: From Today's Community Innovation Surveys to Better Surveys Tomorrow*, Paris: Organisation for Economic Co-operation and Development and Eurostat, http://www.oecd.org/dataoecd/37/39/37489901.pdf, document dated 6 September 2006 (accessed 4 February 2010).

OECD (2009), *OECD Reviews of Regional Innovation: 15 Mexican States*, Paris: Organisation for Economic Co-operation and Development.

Pavitt, K. (1984), 'Sectoral patterns of technical change: towards a taxonomy and a theory', *Research Policy*, **13** (6), 343–73.

Phelps, N., R. Fallon and C. Williams (2001), 'Small firms, borrowed size and the urban-rural shift', *Regional Studies*, **35** (7), 613–24.

Piore, M. and C. Sabel (1984), *The Second Industrial Divide: Possibilities for Prosperity*, New York: Basic Books.

Polèse, M. and R. Shearmur (2002), *The Periphery in the Knowledge Economy*, Montréal: INRS-Urbanization Culture et Société, et Moncton: ICRDR.

Porter, M. (2003), 'The economic performance of regions', *Regional Studies*, **37** (6–7), 549–78.

Romer, P. (1989), *Human Capital and Growth: Theory and Evidence*, NBER working paper no. W3173.

Shearmur, R. (2010a), 'Innovation, regions and proximity: from neo-regionalism to spatial analysis', *Regional Studies*, on line http://dx.doi.org/10.1080/00343404.2010.484416.

Shearmur, R. (2010b), 'Space, place and innovation: a distance-based approach', *The Canadian Geographer*, **54** (1), 46–67.

Shearmur, R. and N. Bonnet (2010), *Like Oil and Water? Regional Innovation Policy and Regional Development Policy*, INRS working paper, 2010–02, Montréal: INRS UCS, http://www.inrs-ucs.uquebec.ca/pdf/inedit2010_02.pdf.

Shearmur, R. and M. Polèse (2007), 'Do local factors explain local employment growth?: Evidence from Canada, 1971–2001', *Regional Studies*, **45** (4), 453–71.

Shearmur, R. and D. Doloreux (2009), 'Place, space and distance: towards a geography of knowledge-intensive business services innovation', *Industry and Innovation*, **16** (1), 79–102.

Shearmur, R., P. Apparicio, P. Lizion and M. Polèse (2007), 'Space, time and local employment growth: an application of spatial regression analysis', *Growth and Change*, **38** (4), 697–722.
Simmie, J. (ed.) (2001), *Innovative Cities*, London: Spon Press.
Simon, C. (1998), 'Human capital and metropolitan employment growth', *Journal of Urban Economics*, **43**, 223–43.
Solow, R. (1956), 'A contribution to the theory of economic growth', *Quarterly Journal of Economics*, **70**, 65–94.
Statcan (2006), *Survey of Innovation*, Ottawa: Statistics Canada, http://www.statcan.gc.ca/cgibin/imdb/p2SV.pl?Function=getSurvey&SDDS=4218&lang=en&db=imdb&adm=8&dis=2 (accessed 4 February 2010).
Storper, M. and R. Walker (1989), *The Capitalist Imperative: Territory, Technology and Industrial Growth*, New York: Basil Blackwell.
Tickell, A., E. Sheppard, J. Peck and T. Barnes (2007), 'Methods matter: transformations in economic geography', in A. Tickell, E. Sheppard, J. Peck and T. Barnes (eds), *Politics and Practice in Economic Geography*, Los Angeles: Sage, pp. 1–24.
Torre, A. (2008), 'On the role played by temporary geographical proximity in knowledge transmission', *Regional Studies*, **42** (6), 869–89.
von Hippel, E. (1988), *The Sources of Innovation*, New York: Oxford University Press.

8. Innovation and creative leadership in local government

Jan Ole Vanebo and Alex Murdock

INTRODUCTION

This chapter explores the evolution of knowledge management through the example of a Norwegian municipality. It also examines how the leadership of the municipality develops a wider awareness and creativity both within the staff of the municipality and through key stakeholders. The knowledge economy has been conventionally viewed in terms of commercial undertakings and there has been a considerable interest in 'spillovers' from individual enterprises to the wider economy (Raspe and Oort, 2008). Such 'spillovers' are seen in terms of factors such as geographical and other proximities. In the context of Haram, the 'spillovers' are geographically mediated by the nature of local government. However, the key industry stakeholders are often global enterprises which are not so bounded. We also note the importance of the global context and the development of the knowledge economy for the setting of this municipality. Jensen (2008) notes some key factors in the development of a 'knowledge society' which we regard as significant in this context. He notes increasing global competition, developments in information and communication technology (the key aspect of more highly skilled labor) and finally the role of the state. In the Haram setting these all are significant factors. We illustrate through the response of the leadership of the commune how this demonstrates the evolution of a knowledge-based economy where the municipality has taken a proactive role with key stakeholders.

In this chapter, we link and combine the organizational knowledge approach to the concept of creative leadership introduced by Philip Selznick (1957) in his essay *Leadership in Administration*, together with Johnsen's (2002) leadership roles approach and Busch et al.'s (2007) strategic change management approach. The creative leadership role has two aspects. First, there is what Selznick called 'institutional embodiment of purpose'. Secondly, creativity is exercised by strategic and tactical planning, that is, analyzing the environment to determine how best to use the existing resources and capabilities of the organization. As stated by Selznick (1957, p. 152), 'the art of the creative leader is the art of institution-building, the reworking of human and technological materials to fashion an organism that embodies new and enduring values'.

Organizational knowledge exists when individuals in an organization share a set of beliefs about causal relationships that enable them to work together. Learning is the process that results in a change in knowledge, that is, change in beliefs about causal relationships in the world and our organization. Learning results in new or modified interpretive frameworks for making sense of the world and taking action in it. We have conducted a longitudinal case study in a Norwegian municipality, Haram, with the purpose of analyzing and conceptualizing managerial activities (leadership roles) concentrating on changing the strategic orientation and the basic character of the actual case municipality. The research questions which we set out to explore therefore are as follows:

1. What is the nature of creative leadership?
2. How important is such leadership in solving the problems in Haram?
3. What is the linkage between such creative leadership and the knowledge economy?

Our approach combines two contrasting methods of creative leadership in the force field between the modernistic (instrumental) view and the social constructive (institutional, interpretative) view. This position is named 'pragmatic institutionalism' by the Norwegian neo-institutionalist Røvik (2008). We also draw on Sanchez (2001) and his five learning cycles and emergence and integration of new knowledge.

The five cycles represent the basic processes through which an organization receives, evaluates, absorbs (or rejects), and deploys new knowledge. Similar to Barnard (1938), Selznick (1957), Johnsen (2002) and Busch et al. (2007), our point of departure is that creative leadership in strategic organizational change, and in particular, the role of leadership in re-orientations is of vital importance. Given organization and individual inertia, re-orientations cannot be initiated or implemented without sustained action by the organization's leadership. A key challenge for executives facing turbulent environments and urging needs for innovations and improvements then is to learn how to effectively initiate, lead and manage re-orientations. This is getting more and more important as the knowledge society evolves.

INNOVATION IN THE PUBLIC SECTOR

During the 1980s a new concept of the state emerged. This is often associated with the emergence of the concept of new public management (NPM) (Hood, 1991). The state should not only administrate, but create public value and facilitate sustainable development. An important aim of creating public value (Moore, 1995) is to focus attention on initiative and innovation as key components of 'strategic management'. To remain efficient, effective and responsive, government managers would have to innovate. Such innovations could increase public value in public organizations in at least three different ways, consisting of improving regulatory functions, service provisions and policy making.

Hartley (2005) emphasizes that innovation is not just a new idea but a new practice. She introduces three approaches to innovation in the public sector in the post-war period and identifies and analyses their implications for policy makers, managers and citizens. We draw on Hartley in illustrating (Table 8.1) three types of innovation and improvement in different concepts of governance and public management and consequences for the role of public managers.

In the traditional public administration paradigm the value focus is on legitimacy and accountability. Hierarchy is the governance structure and authority the coordinating mechanism. The logic is characterized by a stable logic of appropriateness (March and Olsen, 1989), and the dominant reform strategy is asserted by Pollitt and Bouckaert (2004) as the maintain strategy. The content of the maintain strategy is: lightening the existing bureaucracy, deregulation and streamlining and saving money by tightening budgets and financial management. Hartley (2005) names the policy maker, commander. The roles of the public managers are clerks or martyrs. The role of policy maker is to act as commander – creating legislation and then support for whole-scale changes, while assuming that the detailed work of implementation will be carried out by officials. These public managers, working within the organizational form of bureaucracy, act either as 'clerks' (impassive officials implementing political will) or as 'martyrs' (holding private views about the wisdom of needing action but continuing to implement political decisions without comment).

A different approach for reform strategy and innovation is seen in the approach known as new public management (NPM) and has developed since the 1980s in the UK. The value focus is on efficiency and responsiveness. Market is a dominant governance structure and price is the coordinating mechanism. The logic is characterized by a competitive logic of instrumentalism. The reform strategies are modernizing and marketizing. The modernizing strategy emphasizes strategic planning, budget reforms, managerial modernization and participating modernization (Pollitt and Bouckaert, 2004). Marketizing is about competition, market and market-like principles reducing the distinctiveness of the public sector vis-à-vis the private sector. The innovations arising through this approach focus particularly on organizational forms and processes such as executive agencies in central government, the purchase–provider splits seen in health, education and local government, and a 'customer' focus. Change initiatives in Haram are obviously inspired and influenced by NPM. The managerial focus of NPM reduces the role of the politicians to 'commissioners' of services

Table 8.1 Paradigms of public value creation

		Paradigm I: traditional public administration	Paradigm II: new public management	Paradigm III: network governance and partnership
Context	Value focus	*Legitimacy Accountability*	*Efficiency Responsiveness*	*Creativity Entrepreneurship*
	Governance structure	*Hierarchy (authority)*	*Market (price)*	*Network (trust)*
	Logic	*Stable: logic of appropriateness*	*Competitive: logic of instrumentalism*	*Continuously changing logic of innovation and improvement*
Reform strategy		*Maintain*	*Modernize, marketize*	*Mobilize*
Role of public policy maker		*Commanders*	*Announcers/ commissioners*	*Statesman*
Role of the public managers		*Clerks and martyrs*	*Businessmen*	*Leaders as interpreters – institutional entrepreneurs, creative role players*
Innovation		Type 1 Better methods for performing core function: regulation and service provision	Type 2 Performance improvement in service provision-adaptation and customization of basic operational procedures to meet the demands of varied circumstances and clients	Type 3 New uses of organizational capabilities in policy making to accomplish current and new mission

or 'announcers' of change (Pollitt and Bouckaert, 2004). Mandarin officials have been transmuted into public managers as efficiency maximizers, seeking innovations to improve the quasi-market and the quality of service 'delivery'. The public increasingly take on customer roles which give them a voice, as users, in service scope and content (Hartley, 2005).

Innovation should materialize in performance improvement, adaptation and customization of basic operational procedures to meet the demands of varied circumstances and clients. In the network governance and partnership paradigm, the value focus is on creativity and entrepreneurship. The governance structure is a cooperative network, and trust, the dominant coordinating mechanism. Reform strategy under networked governance is characterized by mobilization and cooperation. Partners shape common missions and strategies and the leadership role of policy makers is to facilitate the translation of new ideas into new forms of action. The role of public managers is to nurture innovation as they become explorers commissioned by society to search for public value. In undertaking the search, managers are expected to use their initiative and imagination. But they are also expected to be responsive to more or less constant political guidance and feedback (Moore 1997). We are arguing that they become creative role players, and

> innovation is focusing on new uses of the organizational capabilities by introducing new products and services that can be used to deal with different parts of their current mission, or even meet a need that is outside their current mission. This could be viewed as a strategic innovation insofar as it repositions the organization in its environment by changing the set of functions it performs and services it provides. (Moore, 1997, p. 299)

This is more or less the recipe for the strategic change initiatives in the Haram case.

Summarizing this in respect of knowledge creation and creativity we note that traditional public administration puts less store in both leadership and the creation of knowledge through crossing (or removing) boundaries between the state and the non-state. It is process-focused and takes its direction from elected politicians. The concept is more of the public servant than the 'public leader'. Innovation has a stress upon methods and improvements in conventional service provision. New public management brought with it the concept of efficiency and the use of market (or quasi-market) mechanisms to encourage this. The focus moved towards performance measurement but with retained thought that the direction was political and there were clear boundaries between actors (politicians, managers and citizens). Innovation was seen in terms of developing more consumer-oriented services using more business approaches.

Network governance and partnership offers a different paradigm. Here the focus is not so much on process (legitimacy) or efficiency (customer) but rather on creativity with a sense that leadership is a more important factor than position. Leadership is not seen as anchored in particular roles (such as mayor or CEO) but is open to innovation and to new ideas. Some have described this as 'new public leadership' (Brookes and Grint, 2010). Here we use this to focus upon both creative leadership and knowledge creation using the Haram example.

CREATIVE LEADERSHIP

Introduction: Definition

Drawing on Johnsen (2002) we define creative leadership as follows: 'Creative leadership entails generating the total of the capabilities (knowledge, competences, resources and power) which are necessary and adequate in order to create public value and accomplish the mission'. This approach to creative leadership is based on four cornerstones of strategic analysis and synthesis: dynamic, systemic, cognitive and holistic views of organizations.

The starting point is strategic analytical leadership and the aim of the analysis is to go through the following model to get relevant, necessary and sufficient information to answer four key questions (as referenced in Table 8.2).

- Who are we?
- What is our mission (purpose)?
- What is our vision?
- Which strategy leads us from the present mission to the realization of our vision?

Innovation and creative leadership is about adjusting and developing the organization to improve efficiency and effectiveness. The vital creative leadership roles are the stakeholder-analyst, entrepreneur, strategist and statesman. The stakeholder-analyst must penetrate into the core of the firm, and then onwards to identify its limits. Table 8.2 could also express questions such as what is our mission today and tomorrow and how do we realize it? It queries the 'we',

Table 8.2 Creative leadership analysis

	Retrospect	Prospect
General view	*We* Dominating coalition (stakeholders) and current mission, management system, transformation system and behavior system	*Vision* Imagination Desired future to enable accomplishment of future mission
Insight	*Mission* Present strategic ends and means – strategic position	*Strategy* Implementation and integration Who does what, when? Sequence of development projects

the strategic position, the 'mission', the desired future strategic position, the 'vision', and finally how can we advance from present strategic position into a new strategy. The entrepreneur role is characterized by having a development objective and employing search-learning processes as means of solving problems. The primary objective is to contribute to the further development of the company's mission, purpose or 'raison d'être', that is, those capabilities that form the basis for the firm's existence. The entrepreneur is a dynamic person who always considers how to:

- realize new initiatives;
- serve customer groups better;
- raise money and resources;
- remove obstacles on the way to ensuring the realization of the project.

The objective of the strategist is also to actively participate in changing the strategic logic and shifting the basic foundation for the firm's existence. As the strategist adopts a behavior-oriented view of the firm, he/she is primarily concerned with actively changing the human behavior. It may be deduced from this that the strategist can take initiative to initiate projects of change within the ranges of systems contained in the firm: the coalition system, the behavior system, the transformation system and the management system. Evidently the statesman role is the most extensive of all the management roles, as the work of the statesman is a combination of managing knowledge (what, why, how) into competence and implementing and integrating the new knowledge into improved practice. The statesman role has three objectives:

- to develop a realistic vision reflecting the organization's desired future position;
- to develop the system's management process and managerial capability;
- to observe how the organization's strategic profile develops/evolves compared to that of a competitor and thus the statesman must continually analyse and identify the strategic aspects of competition, distinct competence and strategic position.

INTEGRATED CREATIVE LEADERSHIP

Creative leadership is an integration of strategic analytical, innovative and political leadership and consequently the four creative leadership roles: the stakeholder analyst, entrepreneur, strategist and statesman. When organizations develop into institutions, they develop an identity and distinct competence and the relation of leadership to organizational character may be more closely explored if we examine some of the key tasks that leaders are called to perform (Selznick, 1957):

1. the definition of institutional mission and role;
2. the institutional embodiment of purpose;
3. the defense of institutional integrity;
4. the ordering of internal conflict.

Selznick (1957, p. 153) draws a distinction between administrative management and institutional leadership: 'If one of the great functions of administration is the exertion of cohesive force in the direction of institutional security, another great function is the creation of conditions that will make possible in the future what is excluded in the present.' This requires a strategy of change that looks to the attainment of new capabilities more nearly fulfilling the truly felt needs and aspiration of the institution. According to Selznick (1957) the executive becomes a statesman as he makes the transition from administrative management to institutional leadership. The statesman role is further developed by Johnsen. According to Johnsen (2002) the statesman works with the development of the overall system. This managerial role is concerned with converting the present strategic balance into a new and more appropriate one. The statesman function entails strategic analysis and synthesis (the stakeholder analyst role), innovative leadership (the entrepreneur role) and political leadership (the strategist role). So the statesman should be capable of handling the interplay between strategic development processes, innovative processes and political processes. The specific tasks of the statesman are:

- to produce the concrete, mutual strategic ends and means of the coalition;
- to identify the organization's strategic developmental phase, and the concrete forces that can be mobilized by individual stakeholder groups as well as the coalition as a whole for strategic development activities.

The statesman's problem-solving methodology is a combination of three different strategic leadership processes: strategic analytical leadership, strategic innovative leadership and strategic political leadership. In strategic analytical leadership the following question is asked and should be answered: 'What is the meaning of the firm/organization/institution?', or, 'What is our mission today and tomorrow, and how do we realize it?' By asking this question the analyst queries the present strategic objectives (the mission), the future strategic objectives (the vision), what we can do today to fortify our existence-mission, and how we can turn our present mission into our future mission? The core of the system is explored as well as its boundaries. The analyst ponders how to move them, in what order, when and what the stakes should be.

The statesman role in local government implies institutional capacity building and institutional capital in the context of local governance. Healey et al. (1999) identify institutional capacity building as the development of knowledge resources, relational resources and mobilization capacity. Inspired by Amin and Thrift (1995), they are concerned with how local governance could get to play a role in reducing vulnerability of local economies, societies and environment to damaging external pressures, while at the same time promoting local economic health and quality of life. Localities which can deliver this beneficial nexus are characterized by four factors: a plethora of civic associations, a high level of interaction between social groups, coalitions which crossed individual interests, and a strong sense of common purpose. These four factors will generate a quality of 'institutional thickness' or richness, within which firms and households are embedded, and an institutional capability to mobilize to sustain supportive conditions for both. Amin and Thrift have developed the dimensions of these four factors into six qualities which make up a 'territorialized system': the persistence of local institutions, a deepening 'archive' of commonly held knowledge (formal and tacit), institutional flexibility (the ability of organizations to change), high innovation capacity, capacity to develop relations and trust and reciprocity, and a sense of a widely held common project.

In summary, one may say that creative leadership (the four leadership roles) is promoting innovation, efficiency and effectiveness through a combination of three different management processes: strategic management processes, innovative management processes and political management processes. Strategic management (stakeholder analyst) relates to an initial process where individuals and local environments continuously take on new projects with the purpose of shifting the organization's 'raison d'être'. In this process both individual and collective learning are realized.

CREATIVE LEADERSHIP IN PRACTICE: THE HARAM CASE

A Brief Summary of Haram

Haram commune has a population of 8750 (2005) and is one of the 36 communes in the county of Møre and Romsdal. In 2007 the Commune employed 550 staff with a total budget of some 500 million Norwegian Kr. (about €62 million). Many of the staff were part-time employees.

The traditional responsibility of such a commune is service delivery. This includes primary schools, kindergartens, primary health care, care for the elderly, social security and maintenance of local roads. Haram also had responsibility for local planning, permits for new enterprises and buildings, and provision of information (libraries).

In Haram no single political party had a majority in the commune and party politics were perhaps less important at the local than at the national level. There was a higher than average proportion of women politicians in Haram and a general agreement on what were the challenges. The actual political make-up meant that the largest party (Senterpartiet) was one that had traditionally represented rural and agricultural interests.

Haram's Mission and Vision

Haram's mission is to a certain degree defined by law: regulation and service provision.

The third element of the mission (Table 8.3) is discretionary and dependent on the will and ability of the municipality, namely, policy making or community development.

Haram's vision is 'to be the leading maritime industrial municipality in Norway'. This vision is discretionary, ambitious and outside the current mission.

Table 8.3 Paradigms of value creation – Haram

		Paradigms of public value creation		
		PA	NPM	Network/partnership
Mission/ tasks of local government and appropriate paradigm(s) of public value creation	Regulatory role	×		
	Service delivery	×	×	×
	Discretionary outside the current mission Vision of community development			× Leading maritime industrial municipality

Haram Situation

Haram is an area on the coast of Norway. The basic level of local government in Norway is called the municipality. In 1999, the municipality produced a strategic plan which was the first step in developing a concept of governance to promote community development as well as the provision of statutory services. In 2003, Haram set up a new management team which was internally recruited and trained. In 2005, a program linked to a nearby university was proposed with the suggestion that this could enable the municipality managers to engage in work to develop aspects of the new municipality plan.

Haram municipality faced two particular challenges. The first was demographic in that the younger working population was drawn away to large cities like Oslo, where there was a greater diversity of employment opportunities. The second was educational in that the level of higher education in Haram was lower than that of the surrounding area, which in turn was lower than that of the average for Norway.

This led to a recognition that the problem would have to be confronted using novel paradigms and finding innovative and creative solutions. The key issues of migration and a potential change in the workforce requirements of local industry were not amenable to solution through improvements – or even innovations – in public service delivery. Rather, the key leaders of the municipality would have to develop new ways of thinking. A key element in recognizing this was to first accept that they needed to work together to find these new paradigms and solutions. The municipality solution was to develop a Master's program with a nearby university, involving consultants (Pricewaterhouse-Coopers) to enable the managers and politicians of Haram to:

- be able to develop a constructive dialogue with the business community;
- construct a second generation municipality plan to cover the period 2006–16.

In early 2006, a series of whole-day seminars were held for the politicians and key stakeholders. The seminars generated key inputs to the final plan and there was active engagement of key stakeholders from the business community, civil society, local media and university and research institutions. The strategic challenges of the community were highlighted through identifying:

- the municipality as an organization (political-administrative system);
- the municipality as a local society.

The New Community Plan

The four key elements of the new community plan (2006–16) were:

1. Haram and the environment;
2. business, competences and value creation;

3. a strong and aligned organization as an organizational actor in society;
4. Haram as a decent place for living, working and leisure activities.

The seminars referred to previously were essential in the process. Involvement of politicians was critical. The final decision was made in 2007. The objectives for the plan process were:

- to create a realistic starting point for realization and implementation of the community plan strategies and accomplishment of the vision;
- to establish commitment among key stakeholders to a common vision and future purpose;
- to have a common understanding of strategic challenges and the management discourse;
- to create a joint approach.

Two aspects were highlighted. The municipality as an organizational actor is the legitimate departure for the development of local society. Traditionally, it has been a focus to organize the community activity around the administrative work and service delivery. The society development has multiple dimensions and is interested in identifying interests from all relevant key stakeholders. It is a challenge to join all of the local society in a mobilization process to create growth, competitive forces and a sincere commitment to develop a local community which is favorable for a decent life.

THE FRAMEWORK OF CREATIVE LEADERSHIP

Based on our theoretical approach and empirical observations from Haram, we have analysed creative leadership. We have approached this using a framework to examine change. Strategic change comprises strategic analysis, political and managerial processes and innovative managerial processes, which again have four basic elements or strategic leverages: mobilization of the coalition system, strengthening the management system, renewing the behavior system and restructuring the transformation system.

Mobilization is about involving key stakeholders and making the organization (or strategic unit) sensitive to strategic change. The main task is mobilization of the key stakeholders of the coalition system, and includes efforts in developing political support by assessing change agent power, identifying key stakeholders and influencing stakeholders. Focus will be on strategic analytic and strategic political leadership – which means the stakeholder-analyst and statesman's role. This implies revealing discrepancies between current and desired states, and creating credible expectations for the change through empathy and support, communication and participation and involvement. It is necessary to work out a vision statement that includes all or some of the following elements that can be communicated to organization members: mission, valued outcomes or description of desired future states and valued features. This element of the desired future state specifies what the organization should look like to achieve the valued outcomes. These four types of creative activities will affect all three types of innovation. They consequently demand different leadership roles to promote the intended and necessary improvements.

Strengthening the management system implies reconfiguration of the management system and to a large extent, a power issue and question of structural design (describing formal management roles and responsibility), management processes and relationships between internal and external actors. Focus will be on political processes – the statesman role. Because organizational transitions tend to be ambiguous and need direction, special structures for managing the change process need to be created. These management structures should include people who have the power to mobilize resources to promote change, the respect of the existing leadership and advocates of change, and the interpersonal and political skills to guide the change process. There are three potential leverages for the extension of leadership: the senior team, broader senior management and the development of leadership and managerial capabilities throughout the organization. Three sets of leadership activities are viewed as important and critical during times of strategic organizational change. These activities are described as charismatic, instrumental and institutional.

Renewing the behavior system is about aligning the formal value platform (what we should do) with the real organizational culture (what we actually do). All four leadership roles are to a various degree needed.

Restructuring the transformation system is heavily dependent on strategic, innovative leadership and the entrepreneurial leadership role. The first stage is to construct an economic model of the production process in order to reduce the production and transaction costs. The next step is the alignment of the physical infrastructure of the organization to ensure that it fits with the organizations' strategic orientation (mission and vision). Technology and the organization of work (work architecture) have to be combined in a way that secures flexibility and efficiency in the production process. By using buffering strategies (administrative and institutional) the technical core may operate almost as a closed system free from disturbances, and so the function of the transformation system can be improved.

Haram ended up with five prioritized themes in their strategy-focused change program:

1. the reorganization;
2. the new management information system (balanced scorecard);
3. creation of a new value platform;
4. cooperation;
5. management (leadership).

These issues may be analysed as elements of a creative and innovative leadership process.

Creative and Innovative Leadership: Outcome

Innovation and documentation are to some extent uneasy bedfellows. Innovation often runs ahead of the documentation of changes and impacts. Managers actively used the approaches they learned and were able to also bring these approaches to their own staff. In the balanced scorecard, the customer service and learning and growth aspects provided a useful and important focus for realigning the municipality, with finance and internal processes (the other aspects used in the scorecard) being more familiar public management territory. Balanced scorecards were successfully set up and used in all 12 of the units of the municipality. Its adoption has led to a wider appreciation by staff of the need to realign the municipality. It has increased the managers' focus on service quality and cost-effectiveness. Many indicators and benchmarks have been achieved and compared with national data. Consequently, the focus has been targeting a better use of economic and human resources. Also, for the managers the focal points have been to develop system-oriented solutions and to work optimally with resources. Through such measurement the politicians were made aware of progress and achievement. The commitment of politicians to the development of an MPA program stands out as a key innovation and a means of fostering creativity. Given the small size of the municipality, putting the whole management group through such a program was a major investment.

The documented effects are as follows.

Renewal of the behavioral system

The management team and employees were more empowered and responsive. Four hundred out of 600 employees participated in the training program from 2004 until 2006. They came from different units and were organized into cross functional groups with the aim to discuss and come up with solutions to concrete operational tasks, such as an improved service delivery. The main focus was also on the first generation of the value platform which was revised as a consequence to attain ownership, acceptance and commitment to the basic values.

The impact of knowledge creation has been demonstrated in further developments in Haram where there have been exchanges between the municipality and local employers. This has been associated with a number of key innovations and knowledge developments. For example, the local school has developed further links with employers and a teacher has been attached to a local business innovation center. Imaginative outreach work has been developed for the minority communities in Haram (of which there are many). This has involved a two-way exchange of culture and awareness which has involved public sector, private sector and community stakeholders. Some key partnerships have emerged to develop infrastructure in the area. The public transport system tends to draw schoolchildren away from the local school towards Aalesund, the nearest large city. The school brought local industry stakeholders on board to advise and this has led to a much more effective campaign to restructure the bus transport system to favor the school. Business and marketing expertise is being brought into the school and in turn the school is seeking to expand to address the needs of local industry and employers.

An interesting development arose when one of the main private sector employers was taken over by a Korean company. In a follow-up visit one of the researchers found that the new owners were, if anything, more supportive of the community engagement of the company. Examples were offered of supporting a disabled sports team and the development of further links between the staff of the company and local organizations. In the case of another major company, a manager in the municipality has actually moved to work in the company. Such transfers at a senior level from the public sector to private (manufacturing) companies are not the norm. We would suggest that the creative and knowledge-based approach of Haram had served to relax boundaries and encourage such possibilities.

Restructuring of the transformation system: improved service delivery system

An evaluation was done by a questionnaire to middle managers and other actors in the municipality. The respondents (58) have different positions within the municipality organization. The questions were raised as statements to which the actual questioned actor should respond to indicate to what degree he/she agreed with the statement. The scale was from 1 to 6, and the higher score, the higher the degree of agreement, indicating innovation and improvement.

The questionnaire had five focused areas and the intention was to find indications whether the changes were regarded as functional or inadequate in respect to several units. The total documented score is shown in Table 8.4.

The average score was 4.1, which was on the positive part of the scale. The scores varied for the different units and a preliminary conclusion is that

Table 8.4 Documented score

Unit	Average score
Health care service	5.0
Social service	4.8
Child care – kindergarten	4.6
Primary school	4.4
Technical, operational unit	4.3
Elderly care – unit I	4.2
Elderly care – unit II	4.0
Other units between	Between 3.9 and 3.0
Total – average	4.1

changes implemented must be followed up by new initiatives and momentum. Managers actively used the approaches they had learned and were able to also bring these approaches to their own staff.

Alignment of tasks, knowledge, resources and adapting new standards and procedures

The learning outcome of this program is quite positive for all involved actors, and visible proofs indicate improvements in aligning tasks, knowledge, resources adapting new standards and procedures.

CONCLUSIONS

We started out with three research questions:

1. What is the nature of creative leadership?
2. How important is such leadership in solving the problems in Haram?
3. What is the linkage between such creative leadership and the knowledge economy?

What is the nature of creative leadership? Inspired by Hartley (2005) and Bovard and Löffler (2007) we draw the conclusion that creative leadership in local government is about balancing strategic interests and expectations of citizens, companies and other stakeholders and the capability to supplement the traditional local government paradigm (public administration) and NPM with leadership of networks. The critical challenge is how to develop communities by ensuring policy coherence across organizational and sectorial borders and levels of governments as well over time to keep up the momentum and secure sustainable development. Public management is in flux and public leaders are operating in the context of continuous change. They need to have imagination and be interpreters, institutional entrepreneurs and creative role players to be able to implement and integrate new organizational capabilities, improving practices and service provision.

How important is such leadership in solving the problems in Haram? Creative leadership is a move from traditional resource management to resource and knowledge management. Knowledge management is about creating and using new knowledge and enabling the staff's contribution to all goals of the organization by motivation, recruiting, training and promotion.

The overall dynamics of the creative leadership in Haram can be analysed by the five learning cycles (Sanchez, 2001). New knowledge appeared at all levels and in the organization. Key stakeholders, managers and all the employees were involved and had the opportunity to articulate opinions and assess existing organizational and group knowledge, address and imagine alternative interpretative frameworks, and explore new forms of knowledge that would play an important role in their new alternative interpretative frameworks.

The new knowledge of the different actors could move 'upward' and downward through the five learning cycles in a process of organizational emergence and integration as the new knowledge is evaluated and selected (or rejected) by the organization at large. At the top of the figure, managers play a key knowledge management role by deciding which knowledge will be institutionalized by the organization. New knowledge accepted by management was integrated into the Haram organization by becoming embedded in Haram's control system, services and process architectures, and other formal systems of interactions. Through Haram's system design process, knowledge accepted by management was disseminated 'downward' to groups and individuals, where the knowledge could be applied and exploited through the activities of groups and individuals. As the internal actors (managers, employees and trade unions) were able to work with available organizational knowledge, their sense-making capabilities were activated to test and critique both new and prior organizational knowledge.

What is the linkage between such creative leadership and the knowledge economy? In the globalized knowledge economy competition, competitiveness and competence are closely interlinked. The key lesson we feel is to demonstrate that the challenges of a global economy, an increased need for higher workforce skills and a blurring of the boundaries between public and private sector all serve to enhance the growth of knowledge and creativity. Furthermore, it is possible to show such knowledge-based evolution at a quite local level. Here it is through the commitment of key stakeholders and recognition that – from the perspective of the local municipality – it is not possible to simply rely on traditional forms of public management. Rather, creativity and knowledge is enhanced by a more proactive and adventurous approach which necessitates a new and more open mind set.

The Haram account illustrates how creativity and knowledge can occur in a localized and relatively small municipality but it involves a number of factors in this case. There has to be the will and willingness among the key stakeholders for this to happen. We assert that a key element was the openness of the stakeholders to new ways of thinking and in particular the commitment made by the senior management and political leadership of Haram. However, a number of factors were influential. The very isolation of Haram (it is not linked by rail or by air to the rest of Norway) may have served to bring people closer. The interesting effect of the takeover by a Korean company could perhaps have been anticipated in that the Korean companies are often family based and so could be viewed as supportive of community engagement.

REFERENCES

Amin, A. and N. Thrift (1995), 'Globalisation, "institutional thickness" and local economy', in Healey et al. (ed.), *Managing Cities*, Chichester: John Wiley, pp. 91–108.

Barnard, C. I. (1938), *The Functions of the Executive*, Boston: Harvard University Press.

Bovard, T. and E. Löffler (2007), *Public Management and Governance*, London: Routledge.

Brookes, S. and K. Grint (eds) (2010), *The New Public Leadership Challenge*, London: Palgrave Macmillan.

Busch, T., E. Johnsen, S.J. Valstad and J.O. Vanebo (2007), *Endringsledelse i et Strategisk Perspektiv*, Oslo: Universitetsforlaget.

Hartley, J. (2005), 'Innovation in governance and public services: past and present', *Public Money and Management*, **25** (1), 27–34.

Healey, P., C. de Magalhaes and A. Madanipour (1999), 'Institutional capacity-building, urban planning and urban regeneration projects', paper for special issue of *FUTURA* (Journal of the Finnish Society for Future Studies, No. 3, 1993).

Hood, C. (1991), 'A public management for all seasons', *Public Administration*, **69** (1), 3–19.

Jensen, H.S. (2008), 'Management and learning in the knowledge society', *Journal of Regional Analysis and Policy*, **38** (2), 130–37.

Johnsen, E. (2002), *Managing the Managerial Process*, København/Olso: DJØF/Universitetsforlage AS.

March, J.G. and J.P. Olsen (1989), *Rediscovering Institutions: The Organizational Basis of Politics*, New York: Free Press.

Moore, M.H. (1995), *Creating Public Value*, Cambridge, MA: Harvard Business University Press.

Pollitt, C. and G. Bouckaert (2004), *Public Management Reform*, 2nd edn, Oxford: Oxford University Press.

Raspe, O. and F. von Oort (2008), 'Firm growth and localized knowledge externalities', *Journal of Regional Analysis and Policy*, **38** (2), 100–116.

Røvik, K.A. (2008), *Trender og Translasjoner*, Oslo: Universitetsforlaget.

Sanchez, R. (2001), *Knowledge Management and Organizational Competence*, Oxford: Oxford University Press.

Selznick, P. (1957), *Leadership in Administration*, London: University of California Press.

PART III

Learning and skills

9. How to improve the region's ability to learn: a micro-level model on regional actor's knowledge and (informal) learning processes

Manfred Walser and Roland Scherer

INTRODUCTION

In the last decades we can state a paradigmatic shift in theories on regional development. Different approaches are drawn upon to explain the success of a region extending beyond the quality of classic location factors. One basic approach is of particular relevance in this respect, which centers on the knowledge and competencies of the regional actors and institutions: the 'learning region' approach.

This chapter is about how we can improve a region's ability to learn. The main hypothesis is that if we introduce an in-depth understanding of learning processes on the micro-level of single actor's learning we can gain new insights into regional development approaches and also we can develop tools and strategies to foster learning regions.

This chapter starts with a short overview of actual theories of regional development which show a diversity of approaches existing side by side. It emphasizes the role of learning and knowledge as one pattern to analyse regional development processes. With the second section we start from a micro-behavioral model on the process of learning and distinguish between different kinds of learning and knowledge with respect to their contribution to regional development processes. The third section describes different types of learning and knowledge. It is followed by a case study of a regional development process in the region Walgau (Austria), which uses different measures to improve regional learning. The last section discusses the use of the micro-level model of learning to enhance the methodological kit of regional development and formulates the need for further research.

LEARNING AND KNOWLEDGE AS A PATTERN TO ANALYSE REGIONAL DEVELOPMENT PROCESSES

Regional science deals with questions of how to explain the growth and development of regions and locations on an empirical foundation. Recently, there have been paradigmatic shifts in theories of regional development, from exogenous to endogenous approaches, from location to development approaches, and from factor-oriented approaches to approaches focusing on regional actors and systems (Thierstein and Walser, 2000). Traditional location theories have to be supplemented – but not replaced – by production factors like cluster, milieu and R&D and by process-related factors like learning processes, structures of decision making and of enabling/convenience (Bieger and Scherer, 2003). Today regional theories should more realistically reflect the 'real world' and therefore increasingly dynamic rather than static approaches are used (Capello, 2008).

An overview of regional development theories analyzing regional growth and innovation shows a broad range of scientific approaches dealing with different focal points (see for example MacKinnon et, al., 2002; Moulaert and Sekia, 2003).

- The Californian school of economic geographers stresses the relationship between technical innovation, industrial organization and location to create new industrial spaces. Important factors are spatial proximity to minimize transaction costs for input–output relations as well as the change of organizational structures (see for example Storper and Scott, 1989; Saxenian, 1994). To these traded interdependencies Storper (1997) added the focus on the untraded interdependencies which constitute the local environment with its underlying set of conventions and practice. The industrial district school and the innovative milieu school focus on SME and emphasize the quality of formal and informal social, economic and political relations in a region to foster a long-term economic development. Important factors are the trend from mass production to flexible specialization and the innovative capacity of SMEs belonging to the same local production system as well as the importance of wider social and institutional supports and multiple relationships and the role of dynamic collective learning processes (Piore and Sabel, 1984) based on discussions concerning the Third Italy (Aydalot, 1986; Brusco, 1986; Beccatini, 1981; Camagni, 1991; Capello, 1999; Keeble et al., 1999). There is an overlapping of the Californian school and the Third Italy school, emphasizing the endogenous institutional potential to generate innovation, the important role of the firm's support space, and the relationships which are based on trust and reciprocal cooperation including hybrid modes of organization between cooperation and competition. Both 'schools' stress the relative significance of local linkages vis-à-vis global networks and the role of culture as a vehicle of change.

- The new geographical economics are a kind of 'mainstream economics revisited'. It takes into account the improved transport and communication technology causing non-linear transportation costs and emphasizes the increasing importance of spatial scale and regional advantages resulting from returns to scale (Krugman, 1991; Porter, 1996; Venables, 1996). It concentrates on economies of agglomeration, also partly referring to aspects of an endogenous development. The models dealing with clusters have no coherent approach but they all refer to Marshall's advantages of local spaces and emphasize the important role of local culture and institutions (Moulaert and Sekia, 2003).
- With the upcoming evolutionary and institutional economics new factors influencing the regional development come to the fore. The idea of path dependency means that past decisions and particular actual choices open up new pathways and preclude others and also throw light on the role of institutions and an incremental form of innovation in shaping the economic development trajectories (see for example Nelson and Winter, 1982; Hodgson, 1993; Storper, 1997; Amin, 1999).
- Also the economic sociology added new insights, for example the concept of embeddedness (Granovetter, 1985) or the meaning of trust and mutual understanding for regional development (Amin and Thrift, 1995; Keeble et al., 1999).

In summary, there is no longer one single central, comprehensive theory of regional science. Instead, different parameters are drawn upon to explain the success of a region beyond the classic location factors.

Many of the theories and approaches on regional development emphasize the regional actors' and institution's knowledge, competencies and ability to learn as one important parameter. The important role of space in innovation processes derives from the advantages of geographical, relational and institutional proximity (Capello and Nijkamp, 2009). Actual approaches center on this context, for example:

- the ability of a multitude of actors to interact, to learn collectively and to build a trustworthy and creative atmosphere results in a region's 'constructed comparative advantages' (see Aula and Harmaakorpi (2008) for the strategy of 'regional reputation building');
- the enhancement of a region's social capital resulting in social and economic comparative advantages (Putnam, 1993);

- the enabling of firms to enhance their 'absorptive capacity', which means the firm's ability to value, assimilate, and apply new knowledge (Cohen and Levinthal, 1990).

Such approaches are examples of how to conceptualize the 'self-reinforcing mechanisms of knowledge creation' (Capello and Nijkamp, 2009, p. 5) as an endogenous determinant for regional development and as a part of a region's 'territorial capital' (Camagni, 2009).

The imperative to learn refers to the fact that regions have to adapt to rapidly changing framework conditions at the global, national and regional level and need to reconfigure their structure of economy to maintain their competitiveness. Therefore, one of the emerging concepts of regional development is the concept of 'learning regions' (see Florida, 1995, 2002; Butzin, 1996; Morgan, 1997; Thierstein et al., 2000; Schläger-Zirlik, 2003), which emphasizes the role of knowledge, qualification and competencies as a fundament for the economic prosperity of a region.

The learning economy can be defined as a rapidly changing economy 'where the success of individuals, firms, regions and national economies reflect their capability to learn … and where the rate at which old skills get obsolete and new ones become in demand is high' (Lundvall, 1996, p. 2). Also, innovation can be defined as a localized process of collective learning to combine complementary forms of knowledge which relies on the continuing trust-based participation of firms and other institutions in an exchange of information and ideas (see MacKinnon et al., 2002).

But collective and organizational knowledge always starts with individual's personal knowledge because 'the organization cannot create knowledge on its own without the initiative of the individual and the interaction that takes place within the group' (Nonaka and Takeuchi, 1995, p. 13). Knowledge can be characterized as the non-material infrastructure of a region depending on creative individuals and groups with their location preferences (Johansson and Karlsson, 2009).

For this reason and as a prerequisite of a learning region, the following explanations center on the process of individual knowledge acquisition of regional economic actors, political decision makers, and other actors influencing the regional economy. To carefully analyse the micro-level of single actor's learning may provide an in-depth understanding of the processes enabling a learning region and of the tools supporting such processes. Capello (2008) asks the question how people actually learn the most crucial aspect of the innovation process, which needs further investigation of cognitive processes in a regional context (Capello and Nijkamp, 2009; see also Cappellin, 2009). The following model may add some considerations to this investigation.

A MICRO-LEVEL MODEL OF LEARNING AND KNOWLEDGE

The meaning of 'learning' is continuously to adapt a system to its changing environment. In this selective and subjective process the individual transforms personal experiences and constructs knowledge. It is the personal background, view of the world, and patterns of interpretation which determine the choice, acceptance and processing of new information. This results in an individually constructed portfolio of knowledge (Dohmen, 2001). Learning in this sense doesn't include any specific quality, and from an external point of view it also can be destructive or pathologic (Willke, 2004). Learning can be described as the activity of processing environmental stimuli using different filters as the following model shows.

The model distinguishes between data, pure and processed information, knowledge and competencies, describing the activities within the learning process and the filters used for these activities as shown in Figure 9.1 below.

Note from this figure several important components:

1. Data are the raw material of information. They are pure inputs open for translation and mostly exist as a sequence of symbols with a pure syntactic

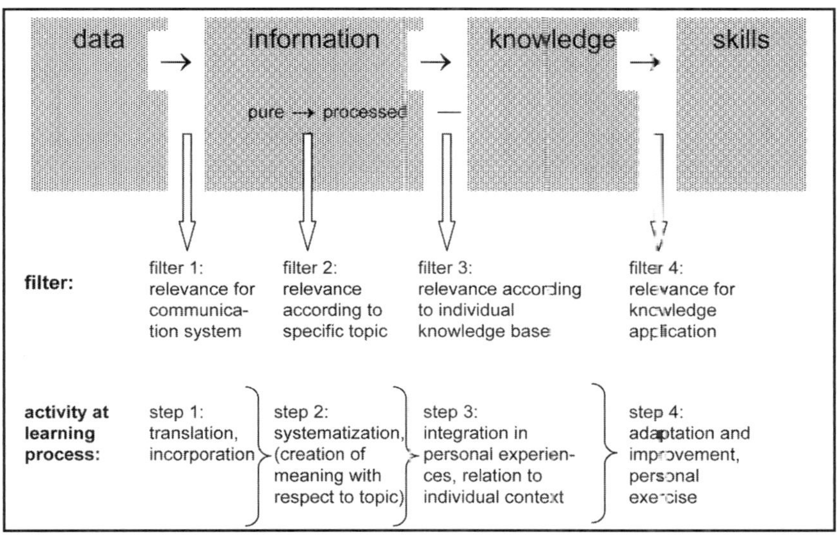

Adapted from Nonaka and Takeuchi (1995) Setzer (2001) and Willke (2004)

Figure 9.1 A model of knowledge and learning

quality: letters, characters, pictures, figures, bits/bytes and so on which have to be processed according to their inbuilt grammar.

2. a. Information is the raw material for knowledge. Pure information is the facts open for interpretation. For example: A region had 35 000 inhabitants in 2009. In 1999 the region had 39 000 inhabitants.
   b. Pure information becomes processed information when it is brought into a certain context and thus gains relevance for a topic. To use the same example: migration, demographic change and regional aging structure are some possible contexts to transform the pure information of inhabitants in 1999 and 2009 into processed information.
3. The step from information to knowledge means that actors integrate the meaningful (processed) information into the context of their personal experiences. (Because it describes a separate step of the learning process, the distinction between information and knowledge shows more clarity as the often used terms 'explicit knowledge' and 'implicit knowledge' which will be introduced in the following section.) Learning persons fit the information into their individual knowledge base and make it useful for them. To again use the example of a region's inhabitants over time: patterns of migration and regional aging structures can have different meanings for different regional scientists depending on their focus of interest or fields of research (labor market issues, care and health services markets, spatial planning, and so on). This active, individual process of acquisition is the bottleneck between the much-discussed 'information overload' and the useful knowledge which is personally available.
4. The last step is taken from knowledge to competency when existing knowledge is put into use. The term 'competency' or 'skills' means having both the authority and the expertise to deal with tasks or problematic situations successfully. For example, this can mean having the ability to successfully deal with the fact of a regional aging population with regard to regional strategy building, whether to act as a researcher, a regional manager or a political decision maker.

Every single step is a learning step on the way to the acquisition of knowledge and skills which are the groundwork for regional development processes. Lundvall (1996, p. 3) described this process with respect to the knowledge economy as follows:

> It is fundamental … that knowledge is something more than information. Information corresponds to the specific elements of knowledge which can be broken down into bits and sent long distance by means of information infrastructures. Therefore, neither is learning, just access to an increasing amount of information. Knowledge includes skills and fundamentally learning is a process of building competencies.

DIFFERENT TYPES OF LEARNING AND KNOWLEDGE

According to the European Commission's definition (2001) we can distinguish between formal, non-formal and informal learning (see Brettschneider, 2004).

- Formal learning usually is provided by an education or training institution, structured (in terms of learning objectives, learning time or learning support) and typically leads to certification. Formal learning is intentional from the learner's perspective.
- Non-formal learning takes place inside or outside of education or training institutions. It doesn't lead to certification but nevertheless it is a kind of systematic learning in terms of learning objectives, learning aid and learning time.
- Informal learning results from daily life activities related to work, social life or leisure. It is not structured (in terms of learning objectives, learning time or learning support) and typically does not lead to certification.

The international discussion on informal learning also deals with related concepts such as 'explicit/implicit learning', 'incidental learning', 'self-directed learning', 'experimental learning' and others (e.g. Straka, 2004; Stangl, 2006). Due to the lack of a standard definition, the empirical research arrives at very different conclusions concerning the importance of informal learning. In different studies, the share of informal learning ranges between 50 percent and 90 percent (see Rohs (2009) for an overview) while the valuation in Northern America is rather more significant and the share ranges for example, around 79 percent (Livingstone, 2000) respectively 70–85 percent (Cross, 2006). The numbers differ according to the research topic which mostly concentrates on learning processes explicitly focusing on job-related skills.

But what is the significance of informal learning with respect to regional development processes? Everyday informal learning is action-oriented and motivated by concrete problem solving (see Bekerman et al., 2006). It is an omnipresent way to interact with one's environment and causes patterns of behavior, routines and values (Trier et al., 2001). In this way informal learning basically shapes the regional milieu and culture in a way Bourdieu (1990) calls 'structure frames habitus frames practice – but the other way round practice can influence habitus can influence structure'. Looking at the different programs of regional development, we have to state that the formalized processes of learning are the predominant ones. Most approaches focus on the role of institutions and courses of (further) education. Processes of informal learning are neither organized in a target-oriented way nor even deliberately reflected. The German network 'Learning Regions', for example, predominantly concentrates on partnerships between educational institutions, economic institutions, and public administrations (see also Scherer and Walser, 2009).

This is astonishing because the tacit knowledge (see Polanyi, 1967) first and foremost is generated implicitly and informally by experiences and practice and needs a process of externalization (Nonaka and Takeuchi, 1995) to be transferred in explicit and transferable knowledge. Tacit knowledge can be stated as one of the most crucial impetuses in most of the theories on regional development mentioned above. MacKinnon et al. (2002) summarize four propositions from the scientific literature on learning regions:

1. new agglomerations become the repositories for tacit specialized local knowledge;
2. relational assets or untraded interdependencies are a key source of learning, enabling regions to effectively deal with change processes;
3. tacit knowledge is embedded in production practices and routines of specific firms and working places, generating clusters of related firms for whom spatial proximity is important;
4. collective learning processes which require a certain degree of continuity and trust of the inter-firm relations and networks are beneficial for innovation and will be promoted by spatial proximity.

Whereas a lot of explicit knowledge influenced by modern information and communication technology is not bound to a specific place, the tacit knowledge is bounded to personal experiences and hence localized knowledge. Lundvall (1996) emphasizes the symbiotic relationship between explicit and tacit knowledge, which implies that tacit knowledge and the individual learning of skills will be fundamental for the economic success of regional actors.

HOW TO FOSTER PROCESSES OF INFORMAL LEARNING: THE REGION WALGAU

Knowledge as the non-material infrastructure of a region (see above) possibly can be improved by systematically using tools which support individual learning processes. An ongoing project of applied research and scientific consultation tests a set of different tools to enhance and foster processes of informal learning and in this way to influence the organizational framework of a learning region. The 'Regional Development Process Walgau' is located in the Western part of Austria in the federal state of Vorarlberg near the borders of Germany, Switzerland and the Principality of Liechtenstein.

The region Walgau is an alpine valley and encompasses 21 towns with around 73 000 inhabitants (Figure 9.2). At the bottom of the valley, big enterprises are

Source: VoGIS, www.vorarlberg.at/atlas

Figure 9.2 The region Walgau

located (mostly belonging to the branches of metal and food processing), while the mountain area is characterized by small rural communities with agriculture and small manufacturing and service enterprises.

The 'Regional Development Process Walgau' is aimed at supporting cooperative projects of the 21 local communities and in a second step to work out a regional development concept out of these activities. The whole process should be organized integrating different target groups and the region's citizens. After the first phase of establishing the organizational structures and procedures, the conference of the 21 mayors and the regional representatives in the federal parliament decided to start the project phase with three core areas to be executed with different measures: (1) to enhance the regional identity; (2) to foster the cooperation between local communities; and (3) to learn from each other. From this working plan, the following three measures will be described more in detail with respect to our topic.

1. Network meetings with different thematic groups: this measure is able to characterize the working culture distinguishing the whole process. Network meetings are organized for target groups out of very different

policy fields which are said to have effects on regional development. For this purpose, actors from social, educational, ecological, agricultural and other fields are invited to exchange their experiences. According to the actor's perception of regional problems and their personal commitment, concrete projects can start from such meetings.
2. Working group of the mountain area: the mountain area is confronted with specific problems deriving from the small-scale settlement structure and some infrastructural deficits. These problems include, for example, public traffic, school sites, local suppliers and the better utilization of agriculture and forestry. Local representatives and exponents of the small villages and mountain settlements periodically meet to get some new information and to exchange their experiences. Such meetings had never happened before and the first meeting showed a surprising consensus on the main problems even among agricultural and tourism-oriented villages.
3. Walgau – Wiki: the regional Wiki should act as a repository of regional information and discussion. Additionally (and in conscious separation from the prototype Wikipedia), every dossier is accompanied by a discussion site which allows not only the discussion of the respective article, but also the discussion on regional visions, goals, and interpretations regarding the issue of the article.

Going back to the model of knowledge and learning we can state that the first two tools are aimed to overcome filter 3 (to integrate information into the personal context) as a prerequisite of knowledge application. Regional actors get the opportunity to widen the bottleneck between information and knowledge because they are provided with additional resources: contacts to other regional actors dealing with similar or complementary issues, a well-defined and structured space in terms of time and arena, and additional attention for their concerns within the region and among regional decision makers.

The third tool – the Wiki – is aimed at overcoming the filter (2b): to create meaning of pure information with respect to the regional context. The Wiki started most recently and it is still too early to assess whether such a web 2.0 tool really fits the needs of actors typically getting involved in processes of regional development. At the moment, it has the character of an experiment.

There are some common principles distinguishing all three measures.

- They all emphasize processes of informal learning in the sense of everyday learning which is action-oriented and motivated by concrete problem solving.
- They frame a process which initiates projects starting from the regional actor's needs rather than from scientific analysis, stressing the importance of individual learning processes and personal commitment as a precondition.

- And the measures are oriented towards creating a common 'view of the world' and common patterns of interpretation of many different actors trying to positively influence the regional milieu.

In this way, the approach of the regional development process can be seen as a 'most stringent way to design a process which is aimed at learning and does not anticipate the issues and goals of this learning process' as the scientific coordinator of the Regional Development Process Rheintal in Vorarlberg characterized it. But the process just started in January 2009 and is, therefore, far from being accomplished. It is an attempt to organize a process in terms of the scientific approaches and hypotheses described in previous sections. To estimate the results of these attempts, a proper evaluation has to be supplemented.

CONCLUSION: WHAT DO INDIVIDUAL LEARNING PROCESSES MEAN FOR REGIONAL DEVELOPMENT?

As Capello and Nijkamp (2009) state, micro-behavioral approaches allows a deep understanding of the sources of the endogenous components of regional development. With our model we are able to identify and analyse the crucial steps of learning with respect to tacit knowledge and competencies of regional actors. Keeping in mind the above delineated model of learning and knowledge, especially the advantage of a differentiation between 'information' and 'knowledge' can be seen: we strongly have to concentrate on step 3 – the integration of information into personal experiences with relation to the individual context. We use this approach to design the Regional Development Process Walgau. Additionally the model can be enhanced towards collective learning processes, which means to integrate an additional filter called 'relevance for collective usefulness', resulting in organizational knowledge.

Learning processes are influenced by different factors, for example goals and topics of learning, instruments and guidance, duration, relevance of results and so on. The question remains: what factors influencing processes of learning have to be attributed to the individual motivation and what factors can be attributed to the design of the learning environment and therefore are open for regulation or at least external influences? In particular, informal learning cannot be guided directly because of its informality. But creating adequate framework conditions like cooperations, working structures or arenas for discourses can support and enhance this kind of learning (see above for a rough sketch of the design of such framework conditions). Understanding the mechanisms of the individual and collective process of learning as sketched in our model can help regional actors to more precisely design the framework conditions shaping such processes.

For this we have to take into account some considerations which should be the object of an in-depth research.

- The above-mentioned networking enables processes of learning and is the precondition to build trustful relationships. Networking is influenced and confined by political and cultural framework conditions. Processes of (informal) learning have to exceed the borders between different societal systems, each with its inherent rationality. The design of learning processes depends on the possibility to use interfaces and 'border crossers' between the systems.
- The exchange of information and the explication of knowledge are horizontally as well as vertically integrated, which means that collective learning is not restricted on spatial boundaries. Also, the exchange between different levels using epistemic communities has to be taken into account.
- The equation of 'proximity causing trust results in regional development' is far too simple. Several empirical studies have documented the problem of lock-in effects and sclerotic networks (see for example Grabher, 1993; Boschma, 2005). Processes of learning have to avoid the limitation of similar perceptions, rising routines and have to enable the exchange with actors from outside the network, with their own societal construction of space.

The three points don't cover the whole program of research but can be seen as a good starting point to deepen our understanding of the interrelationship between regional development and the (informal) learning at the single actor's level.

REFERENCES

Amin, A. (1999), 'An institutionalist perspective on regional economic development', *International Journal of Urban and Regional Research*, **23**, 365–378.

Amin, A. and N. Thrift (1995), 'Globalisation, institutional "thickness" and the local economy', in P. Healey, S. Cameron, S. Davoudi, S. Graham and A. Madani-Pour (eds), *Managing Cities: The New Urban Context*, Chichester: Wiley, pp. 91–108.

Aula, P. and V. Harmaakorpi (2008), 'An innovative milieu – a view on regional reputation building case study of the Lahti urban region', *Regional Studies*, **42** (4), 523–38.

Aydalot, P. (1986), *Milieux innovateurs en Europe*, Paris: Groupe de recherche européen sur les milieux innovateurs GREMI.

Becattini, G. (1981), 'Le district industriel: milieu creative', *Espaces et Sociètès*, **66/67**, 147–64.

Bekerman, Z., N.C. Burbules and D. Silberman-Keller (2006), *Learning in Places. The Informal Education Reader*, New York: Peter Lang.

Bieger, T. and R. Scherer (2003), 'Clustering und integratives Standortmanagement – von einem theoretischen Konzept zur konkreten Handlungsstrategie', in R. Scherer and T. Bieger (eds), *Clustering – das Zauberwort der Wirtschaftsförderung*, Bern: Haupt, pp. 9–26.

Boschma, R.A. (2005), 'Proximity and innovation: a critical assessment', *Regional Studies*, **39**, 61–74.

Bourdieu, P. (1990), *The Logic of Practice*, Stanford: Stanford University Press.

Bretschneider, M. (2004), *Non-formales und informelles Lernen im Spiegel bildungspolitischer Dokumente der Europäischen Union*, Bonn: Deutsches Institut für Erwachsenenbildung.

Brusco, S. (1986), 'Small firms and industrial districts: the experience of Italy', in D. Keeble and E. Wever (eds), *New Firms and Regional Development in Europe*, Beckenham, Kent: Croom Helm, pp. 184–202.

Butzin, B. (1996), 'Kreative Milieus als Elemente regionaler Entwicklungsstrategien? Eine kritische Wertung', in J. Maier (ed.), *Die Bedeutung kreativer Milieus für die Landes- und Regionalentwicklung*, Arbeitsmaterialien zur Raumordnung und Landesplanung 153, Bayreuth: Universität Bayreuth, pp. 9–38.

Camagni, R. (1991), *Innovation Networks: Spatial Perspectives*. Edited by R. Camagni on behalf of GREMI (Groupe de recherche europeen sur les milieux innovateurs), London and New York: Belhaven Press.

Camagni, R. (2009), 'Territorial capital and regional development', in R. Capello and P. Nijkamp (eds), *Handbook of Regional Growth and Development Theories*, Cheltenham, UK and Northampton, MA, USA: Edward Elgar Publishing, pp. 118–32.

Cappellin, R. (2009), 'The analysis of regional knowledge networks', in R. Cappellin and R. Wink (eds), *International Knowledge and Innovation Networks. Knowledge Creation and Innovation in Medium-technology Clusters*, Cheltenham, UK and Northampton, MA, USA: Edward Elgar Publishing, pp. 78–185

Capello, R. (1999), 'Spatial transfer of knowledge in high technology milieux: learning versus collective learning processes', *Regional Studies*, **33**, 353–65.

Capello, R. (2008), 'Regional economics in its fifties: recent theoretical directions and future challenges', *Annals of Regional Science*, **42** (4), 747–67.

Capello, R. and P. Nijkamp (2009), 'Introduction: regional growth and development theories in the twenty-first century – recent theoretical advantages and future challenges', in R. Capello and P. Nijkamp (eds), *Handbook of Regional Growth and Development Theories*, Cheltenham, UK and Northampton, MA, USA: Edward Elgar Publishing, pp. 1–15.

Cohen, W. and D.A. Levinthal (1990), 'Absorptive capacity: a new perspective on learning and innovation', *Administrative Science Quarterly*, **35**, 128–52.

Cross, J. (2006), *Informal Learning: Rediscovering the Natural Pathways That Inspire*, San Francisco: Pfeiffer.

Dohmen, G. (2001), *Das informelle Lernen. Die internationale Erschließung einer bisher vernachlässigten Grundform menschlichen Lernens für das lebenslange Lernen aller*, Research report edited by Bundesministerium für Bildung und Forschung, Bonn: BMBF.

European Commission (2001), *Making a European Area of Lifelong Learning a Reality*. Communication from the Commission, Bruxelles: Directorate-General for Education and Culture and Directorate-General for Employment and Social Affairs.

Florida, R. (1995), 'Towards the learning region', *Futures*, **27** (5), 527–36.

Florida, R. (2002), *The Rise of the Creative Class And How It's Transforming Work, Leisure, Community and Everyday Life*, New York: Perseus.

Grabher, G. (1993), 'Wachstums- Koalitionen und Verhinderungs- Allianzen. Entwicklungsimpulse und -blockierungen durch regionale Netzwerke', *Informationen zur Raumentwicklung*, **11**, 20–31.

Granovetter, M. (1985), 'Economic action and social structure: the problem of embeddedness', *American Journal of Sociology*, **91**, 481–510.

Hodgson, G.M. (1993), *Economics and Evolution: Bringing Life Back into Economics*, Cambridge, MA: Polity Press.

Johansson, B. and C. Karlsson (2009), 'Knowledge and regional development', in R. Capelloa and P. Nijkamp (eds), *Handbook of Regional Growth and Development Theories*, Cheltenham, UK and Northampton, MA, USA: Edward Elgar Publishing, pp. 239–55.

Keeble, D., C. Lawson, B. Moore and F. Wilkinson (1999), 'Collective learning processes, networking and "institutional thickness" in the Cambridge region', *Regional Studies*, **33** (4), 319–32.

Krugman, P. (1991), *Geography and Trade*, Cambridge, MA: MIT Press.

Livingstone, D.W. (2009), *Exploring the Icebergs of Adult Learning: Findings of the First Canadian Survey of Informal Learning Practices*, The Research Network on New Approaches to Lifelong Learning, NALL Working Paper 10–2000, http://www.oise.utoronto.ca/depts/sese/csew/nall/res/cjsaem.pdf, (accessed 6 December 2009).

Lundvall, B.-Å. (1996), *The Social Dimension of The Learning Economy*. DRUID Working Paper No. 96-1, Danish Research Unit for Industrial Dynamics: Copenhagen Business School/Aalborg University, April 1996.

MacKinnon, D., A. Cumbers and K. Chapman (2002), 'Learning, innovation and regional development: a critical appraisal of recent debates', *Progress in Human Geography*, **26** (3), 293–311.

Morgan, K. (1997), 'The learning region: Institutions, innovation and regional renewal', *Regional Studies*, **31**, 491–503.

Moulaert, F. and F. Sekia (2003), 'Territorial innovation models: a critical survey', *Regional Studies*, **37** (3), 289–302.

Nelson, R.R. and S.G. Winter (1982), *An Evolutionary Theory of Economic Change*, Cambridge, MA and London: Belknap Press.

Nonaka, I. and H. Takeuchi (1995), *The Knowledge-Creating Company. How Japanese Companies Create the Dynamics of Innovation*, New York: Oxford University Press.

Piore, M.J. and C.F. Sabel (1984), *The Second Industrial Divide: Possibilities For Prosperity*, New York: Basic Books.

Polanyi, M. (1967), *The Tacit Dimension*, New York: Anchor Books.

Porter, M.E. (1996), 'Competitive advantage, agglomeration economies and regional policy', *International Regional Science Review*, **19** (1–2), 85–90.

Putnam, R.D. (1993), *Making Democracy Work*, Princetown, NY: Princetown University Press.

Rohs, M. (2009), 'Quantitäten informellen Lernens', in M. Brodowski, U. Devers-Kanoglu, B. Overwien, M. Rohs, S. Salinger and M. Walser (eds), *Informelles Lernen und Bildung für eine nachhaltige Entwicklung. Beiträge aus Theorie und Praxis*, Schriftenreihe, 'Ökologie und Erziehungswissenschaft', Opladen: Verlag Barbara Budrich, pp. 35–42.

Saxenian, A. (1994), *Regional Advantage: Culture and Competition in Silicon Valley and Route 128*, Cambridge, MA: Harvard University Press.

Scherer, R. and M. Walser (2009), 'Regionalentwicklung, Wissensmanagement und informelles Lernen', in M. Brodowski, U. Devers-Kanoglu, B. Overwien, M. Rohs, S. Salinger and M. Walser (eds), *Informelles Lernen und Bildung für eine nachhaltige Entwicklung. Beiträge aus Theorie und Praxis*, Schriftenreihe 'Ökologie und Erziehungswissenschaft', Opladen: Verlag Barbara Budrich, pp. 217–25.

Schläger-Zirlik, P. (2003), *Der Ansatz der Lernenden Region in der Stadt- und Regionalentwicklung*, Dissertation. Bayreuth: Universität Bayreuth.

Stangl, W. (2006), *Latentes, passives, implizites, inzidentelles oder informelles Lernen*, Arbeitsblätter, 2006, http://arbeitsblaetter.stangl-taller.at/LERNEN/LatentesLernen.shtml (accessed 6 December 2009).

Storper, M. (1997), *The Regional World: Territorial Development in a Global Economy*, London: Guilford Press.

Storper, M. and A.J. Scott (1989), 'The geographical foundations and social regulation of flexible production complexes', in J. Wolch and M. Dear (eds), *The Power of Geography: How Territory Shapes Social Life*, Boston: Unwin Hyman, pp. 208–15.

Straka, G.A. (2004), *Informal Learning: Genealogy, Concepts, Antagonisms and Questions*, ITB – Forschungsberichte 15/2004, Institut Technik und Bildung, Bremen: Universität Bremen.

Thierstein, A. and M. Walser (2000), *Die Nachhaltige Region*. Ein Handlungsmodell, Bern: Haupt.

Thierstein, A., K. Schedler and T. Bieger (2000), *Die Lernende Region*, Bern: Haupt.

Trier, M., T. Hartmann, G. Aulerich, I. Bootz, P. Buggenhagen, S. Busch, K. Gogolek, P. Heller, G. Hesse, H. Scholz and O. Wolle (2001), *Lernen im sozialen Umfeld. Entwicklung individueller Handlungskompetenz. Positionen und Ergebnisse praktischer Projektgestaltung*, QUEM-report Heft 70, Arbeitsgemeinschaft Betriebliche Weiterbildungsforschung e. V (ed.), Projekt Qualifikations–Entwicklungs–Management, Berlin.

Venables, A.J. (1996), 'Equilibrium locations of vertically linked industries', *International Economic Review*, **37**, 341–59.

Willke, H. (2004), *Einführung in das Systemische Wissensmanagement*, Heidelberg: Carl Auer.

10. Managing the new knowledge workers

Kjell-Åge Gotvassli

INTRODUCTION

During recent years two significant changes have taken place in relation to the nature of the critical resources of organizations and who owns and controls them. The critical resources in today's organizations have changed from being of physical and material character, such as capital and production equipment, to so-called knowledge capital or intellectual capital in the form of the skills and attitudes of individual employees. There has also been a shift in power in relation to who possesses and controls the critical resources – from employer to employee (Nordhaug, 1999; Hillestad, 2000). This means that very many organizations face a new type of employee who is very conscious of his own value and who thereby can be seen as more demanding and less loyal to the company. Rohlin et al. (1994) depict these differences by observing that we are moving away from yesterday's 'industrial society' and into today's knowledge society in which learning and skills development are the most important factors. The employees are the primary engine of the business because they are intelligent, creative and full of initiative. But they are also labelled in advance as resistant to the leadership, willful, flighty and self-important – and hysterically unreasonable and demanding when they do not get their own way (Hein and Larsen, 2007).

In this chapter I will examine some of the challenges this poses for the development of a style of personnel leadership that can promote learning and skills development in organizations. There are many challenges in the field of both leadership and personnel management within knowledge-based organizations. Relevant questions include the following.

1. The problem of recruitment – how should skilled workers be recruited and retained? How should work be organized and made appropriate?
2. What are the principal motivational factors?
3. What forms of leadership are attractive?
4. How can skills development be promoted in an organization?

These four questions have been central in the management literature and in this chapter I will try to discuss them with relevance to management and knowledge workers.

KNOWLEDGE-BASED ORGANIZATIONS AND WORKERS

When skills and knowledge are the most important resources of an organization, we often use the term 'knowledge-based organization'. Knowledge-based organizations use skills and knowledge as input and the organization's internal processes are often connected with communication, coordination, collaboration and skills development. Thus, knowledge-related work is often associated with work that lacks routines, is often complex with use of knowledge from many disciplines, involves both implicit and explicit knowledge and has a high intensity of coordination, communication, interaction and knowledge and skills development (Morgan, 1991; Gottschalk, 2004; Nordhaug, 2002).

Newell et al. (2002) states that there are three characteristics of knowledge-based work and knowledge-based organizations: autonomy, multiple geographical 'workplaces' and 'gold-collar workers'. Knowledge-based workers are highly autonomous in carrying out their work and they expect a high degree of individual freedom. They often have a loose connection to a physical workplace because they are out with customers, and they carry out research and development work in many areas or carry out a great deal of creative work. The term 'gold collar worker' was coined by Kelly (1985) and alludes to employees who are intelligent, independent and innovative and who engage themselves in complex problem solving through challenging tasks in which the results can often be difficult to quantify. In this way they are distinguished from white-collar workers and the traditional blue collar workers. The term 'Generation-X workers' has also been used (Colbjørnsen et al., 2001). It is typical that they have a critical relationship to authority, are geared up for change, positive to variety, enjoyment-oriented and impatient.

Hillestad (2002) defines knowledge-based workers as well-qualified workers with personal abilities, knowledge and/or skills which contribute to increasing, realizing and stimulating an organization's wealth creation and working processes. In line with this, Karlsen and Langseth (2005) say that a knowledge-based worker is a worker with a high degree of knowledge capital and a well-developed sense of organizational belonging. They have substantial skills in their area but often feel the need to bring in third-party skills to solve tasks. They are often very conscious that they are sought-after workers and they make considerable demands and expectations of their employers. They are aware that the most important resources are various forms of knowledge and skills, that the latter are primarily connected to people and that the company cannot own this critical resource.

The development that has been sketched here – from an industrial society to a learning society – can mean that the whole of our Western industrial society is in the midst of a massive anti-standardization of the workplace. More and more employees will become what Beck (2002) terms 'multi-activity nomads'.

These viewpoints are based on widespread diagnoses of contemporary society which include theories of the post-industrial society, the information society, the knowledge society and the service society. They move between skills areas and jobs, often deal with several jobs at the same time and in the course of their working life can move between several jobs, temporary positions and various types of self employment. Many describe such employees as prima donnas. They carry out work which requires meta-skills, in other words an overall knowledge-base, acquired through a long period of education or training within the relevant academic discipline, and where the employee uses these meta-skills creatively to solve complex tasks.

It can be questioned if this is an accurate diagnosis. Frisvold and Leiulfsrud (2003) are questioning three problematic elements. First, they question the fruitfulness of the idea that the industrial society has come to an end. The second question regards central assumptions in the debate about flexibility. Third, they question the description of empowered and autonomous employees. Their empirical claim, based on Norwegian data, is that contemporary working life is not predominantly characterized by fragmentation and dehierarchization.

The image I have drawn up of 'the new workers' may easily give the impression of being something universal, lasting and ever-more dominant. It is therefore important that in shaping a personnel policy some moderating reservations are included. Nesheim (2006) points out that although some changes can be detected in workers' preferences, there will be some uncertainty about the extent to which these represent a stable and lasting generational shift or whether these are characteristics that will decrease with age and stage of life. Economic forces will play a part, and the demands of family life and predictable income will have a disciplining influence and tend towards stability. In a more dramatic manner such issues as climate change, terror threats and epidemics will fundamentally shape and change younger workers' perceptions of reality and work-related preferences.

The knowledge-based worker both works and thinks differently from the classical worker in the 'industrial society' and this can offer considerable challenges in the shaping of leadership roles and personnel policy. In popular terminology there is a saying that leading knowledge-based workers is like shepherding the royal hares, leading an organization with only chiefs; or, put more positively, reflective of the notion that free-range hens lay the best eggs. Løwendahl (1999) has taken some of these leadership challenges to their logical conclusions.

- It can be difficult to establish loyalty in organizations that employ many workers on a short-term basis.
- Neutral offices or open-plan solutions produce little sense of belonging and identity.
- Variable working hours and workplaces lead to people not knowing who is whom or who one is likely to meet – what will happen to the apprentice at the pump?
- Technological solutions make it possible not to meet physically and culturally – the social glue disappears.
- Who 'owns' the company when we risk central workers taking their knowledge and customer relations and moving on – establishing their own businesses or moving over to a competitor?
- How can we develop the careers of virtual employees; how can we become familiar with their strengths, potential, talents, wishes and needs?
- When is there need for administrative support services, and where should they be located in the organization?

The characteristics of a knowledge-based organization and its knowledge-based workers bring with them challenges in terms of both leadership and personnel services. How is it possible to recruit and retain the best heads and how should their work be led, organized and made appropriate? The following takes as its starting point the work of Mintzberg (1999), Hillestad (2002), Colbjørnsen et al. (2001), Nesheim (2006), Kirkeby (2004), Hein and Larsen (2007), Newell et al. (2002, pp. 72–92) and others. In summary, these contributions point to four main strategies: best practice, individually designed strategies, the intellectual capital perspective and the significance of the psychological factor.

STRATEGIES FOR MANAGING THE NEW KNOWLEDGE WORKERS

Use of Organizational Recipes

This perspective uses theories, methods and tools developed by others and implemented into one's own organization. These organizational recipes are ready-made solutions for organizational problems and managing knowledge workers. Themes such as team building, skills leadership, the learning organization and the term empowerment are, according to Røvik (1998, pp. 11–42), examples of an institutionalized super-standard. Other examples are total quality leadership and goal management. Within management of the new knowledge workers there are a number of typical recipes, often termed concepts or management tools, which most modern organizations today are obliged to follow.

The post-war period has produced a number of legitimate recipes for solving organizational problems. That such a recipe is institutionalized means that over a period it will be regarded by many people as the right, sensible, effective, modern solution – even the natural way to lead and organize. These recipes vary in terms of how widespread, international and long-lasting they are.

Why do we stick to recipes? The institutionalized recipe is both a solution and an ideal. What, then, is the explanation behind the spread of these internationalized super-standards? The simplest explanation is, of course, that these recipes are genuine and good solutions to the associated problems. They can be associated with terms like effectiveness, profit and progress. But we are often disappointed because there are no simple solutions to complex organizational problems.

We have a use for ideals and authorities who can legitimize the goals for which we strive. Politicians and administrators must appear to be decisive and modern. They achieve legitimacy by using what is modern. This explanation is also partly a consequence of the position of continuous renewal and development – the vision of moving towards that which is better – as something of central value in our society. One last explanation is found in society's wave of individualism. Those recipes that signify freedom, less governmental control or interference and an emphasis on competitiveness have potential to become standard solutions.

Not all such recipes will necessarily be used directly: a degree of editing or adaptation is often necessary before they can be used within an individual organization. This means that recipes can more easily be moved and put into use (Røvik, 1998, pp. 152–70). This type of editing of organizational recipes can occur as a concretization; partly imitation, combination or re-casting. This means that ideals can be put into concrete form in the shape of advice adapted to the individual organization, that one imitates parts of the recipe, couples together different recipes or attempts to re-cast them into a new form – a new variant.

One can of course believe that these recipes accomplish what they promise, in other words that effectiveness, profitability and progress will follow. But it may also be the case that organizations uncritically copy recipes due to the powerful need to acquire legitimacy and to appear modern. The outward justification is very rational – it leads to profitability, and so on. The rational justifications will also be taken for granted: no-one questions them particularly critically.

The development of an institution's distinctive identity, sets of norms and rules that are to apply occurs over a period of time and is, according to Zucker

(1987), the result of external demands and internal needs. Such pressure can in certain circumstances force an organization to select solutions on the basis of criteria other than effectiveness. For instance, it will be important to select legitimate solutions on the basis of external demands. It appears, for instance, to be in 'everyone's' interest to be a learning organization.

The act of selecting solutions with external legitimacy can lead over a period to a high degree of uniformity among organizations in terms of choice of solution. There will be trends and lines of development onto which many jump in order not to be outside. This is in order to ensure survival in relation to significant aspects of the external world.

This kind of likeness is often termed isomorphism. Di Maggio and Powell (1991, p. 67) distinguish between three types of isomorphism: that originating from political influence and the quest for legitimacy, that due to standard patterns of reaction for overcoming uncertainty, and that which has its background in professional demands and expectations. In relation to leadership in knowledge-based organizations, all three types of isomorphism will be significant influences. Organizations select different strategies, methods and tools by which to acquire legitimacy; to emerge from uncertainty and result in strong professional demands and expectations. Under these circumstances the rational is not always in the forefront, even though rational explanations will always be used when the various strategies are being justified.

In summary, we can say that use of organizational recipes can be appealing because they offer simple, ready-made solutions that can be rapidly implemented. The problem is that the solutions are often too generalized and therefore unable to cope with the fact that organizations and knowledge workers are different. Solutions are also chosen primarily on the basis of external legitimacy and not necessarily because they are good solutions in terms of personnel policy. It can, of course, be useful to use ready-made concepts and solutions but a good deal of tailoring is often needed before they can become optimal solutions for an individual organization with individual knowledge workers.

Tailoring Strategies

Hansen et al. (1999) distinguishes between two main strategies for managing knowledge: codification and personalization. In knowledge codification, leadership strategy springs from the assumption that knowledge can be codified and stored (e.g. in various types of written sources and databases) and afterwards distributed and shared through systems via databases. In personalized knowledge, the leadership strategy is formed on the basis that individuals have the greatest role in sharing and developing knowledge. Knowledge is primarily created and developed by the individual and is best shared by working together, with a great deal of face-to-face contact.

From an instrumental perspective, knowledge-based leadership is viewed as a system which rationally leads to solutions and thereby to improvements. From a reflex perspective knowledge is acknowledged as an important resource for business, and the leadership of this knowledge will be a distinctive discipline because many businesses today are more dependent on knowledge than on material resources. Knowledge leadership is based on an analysis of how knowledge as a resource in a business can change leadership and organization. The focus is not only on the leader's activities but on all activities in and across the knowledge-based business. In addition to the decisions taken by the leader, there is also an emphasis on the decisions taken by the individual skilled employees.

In their survey of different organizations, Hansen et al. (1999) show that the organizations that are most successful in their leadership of knowledge resources often choose one dominant strategy rather than attempting to use both strategies at the same time. The choice of what is best depends on, among other things, the type of solutions offered to customers. It can appear that those who offer standardized services to a large number of clients often succeed with a codification strategy, while those who offer individually designed services to a small number of customers often employ a personalized strategy.

We should not read too much into the study of Hansen et al. (1999) since it is based primarily on American consultancy firms, but one point is significant: with respect to leadership of knowledge-based resources there are no general solutions that can simply be implemented in different types of organizations. A successful strategy for the leadership of a knowledge-based organization demands adaptation and individual design, and it is necessary to analyse a number of factors for a choice of effective personnel policies. It also means that the approach is leader-led and often can seem to be from the top down, something that denies staff the initiative and opportunity for collaboration in analysis and implementation.

Intellectual Capital Perspective

Great interest has focused on development of intellectual capital in organizations. The term intellectual capital has been variously defined in the literature. A comparison of various authors indicates that the term is often divided into three and that it comprises the components found in Table 10.1 (Stewart, 1997; Sveiby, 1990; Edvinsson, 2002).

Human capital consists of the skills of individual members of staff. Structural capital includes forms of collaboration, systems, routines and regulations. Relationship capital involves connections and relationships with customers, distributors, media, authorities and other organizations. In addition, it has also become fairly normal to take account of change and innovation capital; the

Table 10.1 Intellectual capital

Stewart	Sveiby	Edvinsson
Human capital	Individual skills	Human capital
Structural capital	Internal structure	Structural capital (organization)
Relationship capital	External structures	Structural capital (customers)

ability to relate to change and to think afresh. Relevant statistics/indicators include extensive use of research and development, skills development, flexibility and renewal.

There exists a considerable degree of agreement about the importance of developing intellectual capital in all organizations. This offers great challenges with respect to personnel policy, in terms of both attracting knowledge-based workers and developing and retaining the intellectual capital in the organization. I will here examine the following factors: recruitment, economic incentives, salary and fringe benefits, work environment, opportunities for development, career opportunities, attractive leadership forms/motivation, flexible organization and identity and reputation.

To recruit an attractive workforce is not easy in a situation in which there is head-hunting competition and where passive methods such as advertising, job databases and so on are inadequate. Many organizations have therefore adopted relatively proactive methods: they operate systematic visits and presentations at educational institutions; they can be guest lecturers for students or actively contribute to the financing of research in areas that are important to the organization. Some larger Norwegian companies have established and financed doctoral programs and endowed chairs. Another method is active contributions to academic networks and conferences in those disciplines within which the organization knows it will need to recruit. Such arenas often offer many opportunities to get to know potential candidates who can be prepared for a prospective job application. The establishment of trainee programs has also become an important opportunity for business to attract young talent.

Attracting a good workforce can, of course, also be achieved by offering economic incentives, salary and various fringe benefits. In Norway the golden halo or joining bonus is currently a rarity (except in sports) but such gestures are not uncommon among consultancy firms abroad. A more normal method of financial compensation is the payment of removal expenses and help with subsidised temporary housing if the change of job also involves moving. Attractive fringe benefits include nursery-school places, medical services, company cars, newspapers/journal subscriptions, telephone provision and various leisure provisions. These fringe benefits serve a double purpose: they attract and retain the workforce and they also relieve the employee and their family in daily chores so that they can direct more energy towards their work (rather than worrying about unstable child-care provision or sitting in the waiting room at the doctor's office). Some companies offer various sorts of washing and cleaning help for some of their employees. Salary is often a well-used and recognized means of recruitment. It is generally agreed that if one is to attract a high-quality workforce, good salary conditions are important.

There is less unanimity about the effects of salary and pay raises in the working environment (Grimsø, 2005). Many people have experienced that the motivational effect of pay raises is relatively transient until the competition catches up or until colleagues who are less well paid learn, demand the same or reduce their effort. If salary is to have a directly motivational effect, it must primarily be coupled to effort and achieved results, and the individual must be able to see clearly that he or she can affect their own result.

The problem with this type of individual reward is of course that it can have a negative impact on workplace cooperation and that it does not take into account the fact that very many results are achieved as the result of joint effort. Various types of group bonuses are therefore beginning to replace a good deal of the individual rewards within organizations.

To offer a working environment which offers opportunities for development both within and outside the job is seen as an important tool to help knowledge-based companies retain and develop the workforce. Learning on the job is often based on experience. This can mean participation in development groups, interdisciplinary teams, complicated projects, and external contact with customers. Opportunities for development outside the workplace traditionally include various types of courses, in-service training and higher education.

Career development has gradually attained significance in modern business. At the same time, the traditional vertical career ladders with a division between subject-based and administrative/leadership careers are clearly making a comeback. Principles include flexible career routes and the opportunity for so-called cross-over careers where an individual can alternate between an academic/technical career and an administrative one. This is widespread within the university/higher-education sector.

In academic terms one can be working to achieve Professor status while at the same time having periods of leadership as Dean or Principal of the institution. In the future we will probably see a strong increase in the kind of career thinking that gives room for vertical and horizontal mobility – motivating many workers. This mobility will not be limited to individual institutions and will be appropriate to different phases of life. Organizational solutions and greater employment flexibility often fall within what is termed borderless careers. This type of open career concept offers more opportunities and will legitimize an individual's career heading in slightly different directions at different times. Such opportunities can be a powerful tool in retaining and developing the good heads.

Various types of coaching, mentoring and guidance are also important tools for career development. In the report *Utvelgelse og utvikling av lederkandidater i kommunal sektor* (Kjerkol et al., 2004), the principal conclusion is that the model that best meets the need within local-authority sectors for development of local-authority leaders increasingly involves coaching, mentoring and guidance.

Results of leadership surveys carried out by the Norwegian Administrative Research Fund (AFF) show that many leaders find leading 'the new workers' to be very challenging. The employees have also expressed a difficult relationship with their leaders – they expect technical expertise and results-focused leaders who can give active comments. In their overview of the theme 'Motivating knowledge-based workers', Hoverak and Ødegård (2006) refer to some of the normal demands and expectations of knowledge-based workers in relation to leadership and organization:

- It is important that the ethical and moral guidelines of the business are so far as possible compatible with one's own values. Status and reputation are closely tied up with one's own work/business.
- Knowledge-based workers are driven by interesting work. The possibilities of development are not primarily connected to the business, and a lack of prospects for future development will lead to thoughts of changing jobs.
- Knowledge-based workers prefer creative tasks and work that demands problem-solving abilities rather than routine work. A lack of communication, slowness and bureaucracy leads to frustration and this will infect the individual's impression of their working environment.
- Knowledge-based workers have a powerful need to work independently and to take part in decision-making processes.
- Knowledge-based workers have a powerful need for recognition of their intellectual and personal contribution to the institution.
- Knowledge-based workers expect their leaders to be active, visible and inspiring.

The leadership role that emerges is more that of enabler than of hierarchical leader. This is a somewhat simplistic conclusion. The need for flexibility and freedom often needs to be tempered by a need for predictability. The background for this is that there is often a lack of leadership in pure project and network organizations. Frequently, a need will arise for clearer formal boundaries to ensure order, define rights and responsibilities and to create a framework for the knowledge-based worker's technical development. We can also not avoid the classic dilemma of how to balance the need for freedom and flexibility with the need for consistency.

Colbjørnsen (2003) highlights the point that the so-called Generation X of new knowledge-based workers is particularly attracted by a flexible organization. Among other things, this can be achieved through diversity, influence, project variation and teamwork. But Colbjørnsen also warns against too strong a focus on flexibility alone without a minimum consistent framework to meet the needs for a sense of security.

Most contemporary organizations claim a good working environment, opportunities for staff development and flexible organization. But is this enough to be successful in a competitive workforce? Have these factors now become self-evident and necessary requirements or are they insufficient to form a basis for a permanent competitive edge in the quest for employees? Nesheim (2006) claims that the above factors must at least be supplemented in an effort to reveal and communicate the organization's purpose, overall mission or social function. If the employees can connect their social identity to the institution, this creates belonging, relationship and motivation. For many, this can be what distinguishes one organization from another in a time when 'everyone' offers teamwork, flexibility and interesting work.

There are two reasons to put forward overall values and social purpose in the quest for a workforce. First, there is great identity-building potential in formulating a vision and overall purpose. On the basis of theories of identity and cultural development (e.g. Schein, 1987) it is reasonable to assume that values connected to an institution's purpose and social function will give greater meaning and have more significance than formulas concerned with being purpose-driven and efficient. Second, it is easier to express the distinctive, special, and unique aspects of an institution than to highlight values at this level. Most companies can be employee-focused, but far fewer can claim to play a consciously social-constructive role – assuming this is formulated in a credible manner.

Organizational identity often consists of who we are and what we stand for. Identity and emotional belonging can thus have an opinion-forming function and in the long run create a greater sense of belonging and loyalty. Identity is primarily directed internally while reputation and image are directed towards others' opinion of the organization. An organization that acquires a positive reputation through branding will gain great advantages that function both internally and externally. Internally, employees have pride in a workplace that has a good external reputation. Externally, it is easier to raise awareness, for instance in relation to recruiting new staff. There are grounds for believing that skilled people will more easily be attracted to companies that are recognized and have a good reputation than to companies with a more fragmentary reputation.

The term employer branding has been adopted for a special area of marketing connected to the challenges of identity and reputation building in the fight for a skilled workforce. Emphasizing the distinctive, permanent and unique aspects of an organization is also important, but this must be authentic, appropriate and credible. If the very demanding expectations are not met, work with identity building and positive reputation will not be successful.

The significance of pride in one's own workplace has begun to be given considerable attention. In their global survey of employee satisfaction, the Great Places to Work Institute uses the following factors as core measurement criteria: trust in the management, pride in the company and a feeling of fellowship. The Norwegian winner in 2007 was the software company SAS Institute. An article in the Norwegian national business newspaper *Dagens Næringsliv* (8 March 2007) reports that one of the factors for satisfaction is that the organization offers courses in meditation and a dedicated meditation room for all employees. A pub, billiards and a fitness room are other constructive ingredients. In addition, staff should have competitive salaries that are appropriate to what they do. There is also extensive use of bonuses at the group level. Another example is what is termed the world's coolest workplace – the headquarters of the search engine Google in Silicon Valley: Googleplex. Here, the facilities include two swimming pools, a sand volleyball court, fitness studios, 11 variously themed cafés that serve free food, kitchen stations with fruit and snacks, a laundry, dry cleaner, billiards tables, football games, video games and a climbing wall. You can also have barber shops, car washes, free medical services, free vaccines, massage, and free lifts. The premises house over a half of Google's 10 000 employees and have a total green environmental profile (*Dagens Næringsliv*, 29 March, 2007) It seems that these types of workplaces will be of special interest to new knowledge workers.

The Significance of the Psychological Factors

As already stated, knowledge-based workers work very independently and expect a high degree of individual freedom. In leadership literature and media in general, these workers are often represented as prima donnas who are difficult to lead. Some other assumptions are that they do not recognize formal authority, do not accept hierarchy and do not accept economic arguments. This is something of an artificial image which rests on the premise that leadership is only a matter of managing the company and leading and distributing work.

To understand managerial challenges one must also understand what are the most important psychological forces of knowledge-based workers. Research in this area points to two central forces – the climactic kick and the feeling of flow (Gardner et al., 2001; Hein and Larsen, 2007).The climactic kick is a powerful psychological condition that arises when an actor takes the public by storm, when the lecturer attains good contact with the audience, when a difficult operation is successful and when a journalist gets his story on the front page. They are moments of stardom in which the individual feels that what he has been working for gives meaning and satisfaction. Such experiences can be rare, so it is often something else that drives these workers on a daily basis: the feeling of being in the flow. The flow is a description of experiencing a good process. This is effort and problem solving which eventually leads to a good result. This demands deep commitment, strong concentration and often, great frustration as well. It is when one is deeply committed and forgets time and place while concentrating on the job at hand. The kick described above will not govern everyday work. The feeling of flow, however, is something that can be experienced sufficiently often that it becomes a focus of direction for daily work. On the road towards this flow, the knowledge-based worker will often have to work through setbacks and frustrations. Managers can easily interpret this as a general dissatisfaction with the availability of resources and with working conditions.

CONCLUSIONS

The promotion of knowledge development and performance in modern organizations is associated with human resources. The growth of a new type of worker presents challenges for traditional leadership functions and personnel policies in today's organizations. In this chapter I have discussed strategies for managing the new knowledge workers. The use of organizational recipes can be tempting because it offers ready-made solutions to complex problems. The problem is that the solutions are often too generalized and therefore unable to cope with the fact that organizations and knowledge workers are different.

With respect to leadership of knowledge-based resources, tailoring strategies stress that there are no general solutions that can be implemented among different types of organizations. A successful leadership strategy for the knowledge-based organization demands adaptation and individual design. It is necessary to analyse a number of factors for a choice of effective personnel policies. I have also discussed the importance of developing the intellectual capital in all organizations.

This strategy involves working with economic incentives, salary and fringe benefits, work environment, opportunities for development, career opportunities, attractive leadership forms/motivation, flexible organization and reputation. Working with these factors provides effective outcomes because organizations that offer incentives, career opportunities and many types of motivation will be of special interest to the personal attributes of the new knowledge workers.

Knowledge-based organizations are demanding to lead. These institutions must be understood within the force field of politics, economics, technical expertise and management. These organizations have a social task and an expectation on the part of their owners to both achieve social goals and be profitable. Knowledge-based organizational roles are affected by technical arguments, academic loyalty and by the distinctive nature of art, medicine and science. The task of knowledge-based workers to strive toward institutional and collective goals can therefore feel insuperable.

Managers must master the technical, economic, personal and artistic aspects. They must, as the organizational researcher James March puts it, be good at plumbing and poetry. One must allow skilled individualists to bloom when this will benefit the organization. Knowledge-based organizations need leaders with authority who are not authoritarian and who accept that their challenge is to lead skilled people, who in their areas of specialization can be academically stronger than their leaders (Jensen, 2008). This form of management is more based on inspiration and guidance than bureaucratic steering and detailed targets. Leaders will often have legitimate power provided by formal roles. In knowledge-based organizations, however, power is located in the informal power relationship inherent to the natural authority of the leader acquired through personal qualities and academic expertise. Leaders in knowledge-based organizations often have to argue on the basis of academic principles rather than bureaucratic arguments. The leadership of knowledge-based organizations is a matter of handling uncertainty, conflict, dualities, dilemmas and paradoxes. There is no recipe for how it should be done. Regardless, it demands a leader with a desire to mobilize people and who can live with conflicting forces. Paradoxes and dilemmas can be constructive forces that often lead to learning and development. As a leader, one must take advantage of these forces – not believe that they can be eliminated.

REFERENCES

Beck, U. (2002), *The Brave New World of Work*, Cambridge: Polity Press.

Colbjørnsen, T. (2003), *Fleksibilitet og forutsigbarhet: arbeid og organisasjoner i endring*, Oslo: Universitetsforlaget.

Colbjørnsen, T., I. Drake and W. Haukedal (2001), *Norske ledere i omskiftelige tider*, Bergen: Fagbokforlaget.

Dagens Næringsliv, 8 March 2007.

Dagens Næringsliv, 29 March 2007.

DiMaggio, P.J. and W.W. Powell (1991), *The New Institutionalism in Organizational Analysis*, Chicago: University of Chicago Press.

Edvinsson, L. (2002), *Corporate Longitude: What You Need to Know to Navigate the Knowledge Economy*, London: Financial Times Prentice Hall.

Frisvold, B.S. and H. Leiulfsrud (2003), 'Samtidsdiagnoser i sosiologien: Forestillinger om det nye arbeidslivet', *Sosiologisk Tidsskrift*, **2**, 154–80.

Gardner, H., M. Csikszentmihalyi and W. Damon (2001), *Good Work. When Excellence and Ethics Meet*, New York: Basic Books.
Gottschalk, P. (2004), *Informasjonsteknologi i Kunnskapsledelse*, Oslo: Universitetsforlaget.
Grimsø, R.E. (2005), *Personaladministrasjon*, Oslo: Cappelen.
Hansen, M.T., N. Nohira and T. Tierny (1999), 'What's your strategy for managing knowledge?', *Harvard Business Review*, March–April, 106–16.
Hein, H.H. and B. Larsen (2007), *De nye Professionelle. Fremtidens Roller for de Velududdannede*, København: Jurist og Økonomforbundets forlag.
Hillestad, T. (2002), Organisering og ledelse av kunnskapsvirksomhet. I Nordhaug, O.: *Kunnskapsledelse. Trender og utfordringer*, 278–95. Oslo: Universitetsforlaget.
Hoverak, J.G. and S. Ødegård (2006), *Motivasjon av kunnskapsmedarbeidere*, AFF Lerderbulletin, nr. 02–2006, Bergen.
Jensen, H.S. (2008), 'Management and learning in the knowledge society', *The Journal of Regional Analysis and Policy*, **38** (2), 130–37.
Karlsen, J.K. and M. Langseth (2005), *Hvor Skal Jeg Sitte? Om å Lykkes i Det Nye Arbeidslivet*, Oslo: Cappelen.
Kelley, R.E. (1985), *The Gold-Collar Workers: Harnessing the Brainpower of the New Work Force*, Reading, MA: Addison-Wesley.
Kirkeby, O.F. (2004), *Den Nye Lederskap*, København: Børsens Forlag.
Kjerkol, O.M., H. Ness, K-Å. Gotvassli and L.K.L. Kolaas (2004), *Utvelgelse og Utvikling av Lederkandidater i Kommunal Sektor*, Steinkjer: HiNT-rapport 20/2004.
Løwendahl, B.R. (1999), Ledelse med løst inventar: Utfordringer i det 21. århundre. I Nordhaug, O, (red.), Ledelse 2000+, 47–70, Oslo: Cappelen Akademisk Forlag.
Mintzberg, H. (1999), 'Covert leadership: notes on managing professionals', *Harvard Business Review*, November–December 1998, 140–47.
Morgan, G. (1991), *Organisasjonsbilder*, Oslo: Universitetsforlaget.
Nesheim, T. (2006), *Fremtidens Ledelse i Kommunene*, Bergen: Samfunnsforskning.
Newell, S., M. Robertson, H. Scarbrough and J. Swan (2002), *Managing Knowledge Work*, New York: Palgrave.
Nordhaug, O. (1999), *Ledelse 2000+. Trender og utfordringer*, Oslo: Cappelen.
Nordhaug, O. (2002), *Kunnskapsledelse. Trender og utfordringer*, Oslo: Universitetsforlaget.
Rohlin, L., P-H. Skärvad and S.Å. Nilsson (1994), *Strategiskt Ledarskap i Lärsamhället*, Dalby: MiL Publishers.
Røvik, K.A. (1998), *Moderne Organisasjoner. Trender i Organisasjonstenkningen ved Tusenårsskiftet*, Bergen: Fagbokforlaget.
Schein, E.H. (1987), *Organisasjonskultur og Ledelse. Er Kulturendring Mulig?*, Oslo: Mercuri Media Forlag.
Stewart, T.A. (1997), *Intellectual Capital. The New Wealth of Organizations*, New York: Doubleday.
Sveiby, K.E. (1990), *Kunnsakpsledning: 101 råd til Ledare i Kunnskapsintensiva Organisationer*, Stockholm: Affärsvärlden Förlag.
Zucker, L.G. (1987), 'Institutional theories of organizations', *Annual Reviews of Sociology*, 1, 465–88.

11. What drives skill-biased regional employment growth in West Germany?

Alexander Cordes

INTRODUCTION

Much is known about how the local economic structure shapes employment growth. Industrial clusters and other forms of sectoral composition are especially well-examined. In contrast, the skill dimension of employment growth is often reduced to human capital indicators either on the left- or the right-hand side of the regression formula. Employment prospects of mid-skilled and unskilled workers are of less interest to the literature but in view of high long-term unemployment the skill dimension is increasingly important for labor market politicians. Furthermore, analyses of high-skilled employment growth neglects regional policy in peripheral regions lacking urban wage premiums or amenities.

There are many factors related to the regional distribution of economic activity and density that lead to the expectation of skill-biased employment growth, meaning that the demand for high- and low-skilled employment or at least the skill level of the content of work varies with the region type. The fact that human capital is increasingly employed in urban areas is not new. However, high unemployment rates in cities lead to the question, where is less skilled labor better off during the development of the local and nationwide knowledge economy? We argue that exogenous developments, most prominently the ICT revolution, increase the possibilities for the firms to exploit cost advantages in more peripheral, rural regions – similar to what can be found at the international level where the transnational division of labor has increased substantially.

The adverse effects of changes in the spatial division of labor on local employment first found attention in the 1970s when industries were forced to deal with the increasing international competition. Spatial implications of industrial restructuring in the UK, for example, were described by Massey (1994), who worked out the movements and closures of firms in the affected industries and the main urban areas.

However, today cities are no longer suffering from restructuring per se. Due to progress in transportation and communication, firms may choose the optimal location for each part of its business. A new spatial division of labor arises which has a clear skill dimension. To analyse the detailed locational factors (density, industry location, industrial specialization) driving the skill-biased regional employment growth is the main goal of this chapter.

THEORY

The theoretical economic literature on regional employment growth is focused on the important role of skilled labor (human capital) to promote the structural change to the knowledge economy. Low-skilled labor is rather of less interest. However, the majority of the literature comes to the conclusion that labor of different skill levels is not evenly distributed and that this incongruence is increasing rather than disappearing. In the following, the main thoughts of different strands are sketched in order to motivate subsequent empirical analysis and provide some explanations for the latter findings.

In the framework of the new economic geography (NEG), two contributions deal with heterogeneous labor: Mori and Turrini (2005) and Amiti and Pissarides (2005). Mori and Turrini assume different preferences of two groups of workers. Skilled workers concentrate in agglomerations because they prefer variety of the goods markets on the one hand and because they are more easily attracted ('imported') from abroad due to relatively lower transport and information costs (compared to other workers). For less skilled workers these costs are relatively high and, additionally, they prefer to avoid the fierce labor market competition in agglomerations. The other cited article by Amiti and Pissarides deals with a labor market pooling effect concerning the local education level. Hence, the incentives to invest in human capital increase in the level of spatial concentration since it is more likely to find an employer demanding this human capital.

As another strand of the literature, Marshallian externalities of labor pooling, knowledge spillover and shared input markets propose concentration of human capital. It is often – more or less explicitly – assumed that less skilled workers benefit from higher productivity (e.g. Fujita and Thise, 2002). Labor pooling and knowledge spillover provide important incentives to migrate to dense areas or accumulate human capital there. A large labor market increases the matching quality and reduces the unemployment risk of horizontally specialized human capital. Specialized knowledge, in turn, diffuses increasingly with the proximity and number of educated workers. Also, recent empirical contributions emphasize the role of employer changes (and thus the labor pooling argument) for knowledge diffusion (Power and Lundmark, 2004; McCann and Simonen, 2005).

Moreover, the human capital demand of the regional economy varies with the industry mix as different industries need different location advantages depending on the type of product. The product life cycle is one example of distinguishing industries' location requirements (Vernon, 1966; Duranton and Puga, 2001). Another criterion is whether the product is standardized or innovation-intensive and specialized (Leamer and Storper, 2001). The general statement of these two criteria is the same: mature industries with standardized products need low production costs which are to be found in the periphery, whereas young high-tech firms require new impulses and partners from urban areas. It is intuitive that the volume of workers in high demand is growing faster where specific location requirements are apparent. Qualification-specific employment growth then simply results from the local industry mix and the degree of how well the industry fits into the region.

Duranton and Puga (2005) go one step further and distinguish the location requirements of the departments, plants and other functions of the value chain and production process. They primarily distinguish skilled headquarter functions such as management or R&D and less skilled production. This notion has been empirically assessed recently by Defever (2006), who basically confirms the different location requirements for skilled and less skilled functions. Further empirical work in the recent past related to this idea has been published by Lovely et al. (2005) and Aarland et al. (2007).

Following Duranton and Puga, there is the possibility of localization effects on both ends of the qualification structure: while the headquarter functions benefit from knowledge spillovers and strategic cooperation in management and R&D in urban areas, production takes place in rural regions with low labor and land costs and – in the ideal case – shared input markets and specialized traffic infrastructure.

The main reason why this latter idea might have become more important in the last decade is the extension of organizational possibilities due to the technological progress in information and communication technologies which facilitates the separation of company units by maintaining communication and information exchange over larger distances (Ota and Fujita, 1993; Duranton and Puga, 2005).

In summary, the local demand for labor is not homogenous; that is, low-skilled and high-skilled employment growth varies with the structural composition of a region. Moreover, all these structural factors (density, industry location, industrial specialization) are main drivers of the knowledge economy. Furthermore, it is well known from the skill-biased technological change approach that demand for low-skilled workers has been decreasing in developed countries for two to three decades. Thus while this empirical fact has been often validated at the aggregate national level, the theoretical considerations described above imply that at the smaller regional level there are positive employment prospects for low-skilled workers. From a political point of view, the prospect of certain favorable labor market conditions for low-skilled workers may provide policy options to tackle their high unemployment risk. So the main research questions are: Do low-skilled and high-skilled benefit from the same structural composition and – if not so – which are the factors promoting skill-specific employment growth? Or, put otherwise: Is the change to a knowledge economy which is associated with agglomeration and knowledge-intensive industries also a promising employment strategy for low-skilled workers?

The empirical analysis is conducted for the West German NUTS3 districts. In the first instance, however, the skill level has to be empirically differentiated. In Germany, the intermediate level is usually defined by the dual system of vocational education (DSVE). It represents a large proportion of three quarters of total employment (subject to social insurance contribution) and consists of tasks and occupations with very different skill requirements. Therefore, we first classify all occupations by three variables related to what is generally meant by 'skill' in order to split up not only employment within the intermediate level, but also concerning the whole workforce.

OCCUPATIONAL SKILL CLASSIFICATION

Data and Cluster Analysis

There is no explicit database of skill characteristics by occupation in Germany. The Microcensus, a yearly survey of 1 percent of the German population, provides information about qualification characteristics and three-digit occupations at the individual level. We made use of the 'scientific use file' version of the data set (SUF) representing a 70 percent sub-sample of the Microcensus data. For regional analysis it was necessary to use a data source other than the Microcensus which allowed only regional analyses at the NUTS2 level. The most recent regional data at the industry-district level were available from the Federal employment statistics for the period from 2000 to 2006. Taking into consideration the timing of the differing data publication, a similar period was chosen from the Microcensus (the years 1998, 2000, 2002 and 2004) to build the occupation database. For consistency between the two data sets, we limited the Microcensus to the non-student civilian labor force in West Germany. The final sample size thus amounted to 156 939 persons in 2004.

Skill characteristics in the questionnaire are about general and vocational education, the individual position in the firm hierarchy and participation in further vocational training. To circumvent the above-mentioned weight problem of the DSVE it was necessary to combine general and vocational education to years of schooling. The position in the firm hierarchy was indexed. It describes the responsibilities of the employee. For the occupation data the averages of these three skill variables were calculated for every employed person by occupation. Finally, the averages of these skill variables over the years 1998–2004 were calculated. While some three-digit occupations were already compiled in the original data, it was also necessary to aggregate some related occupations in those cases where the underlying number of persons was too small. The final number of aggregated occupations amounted to 288. They were treated as the initial cluster objects.

Cluster analysis attempts to determine the natural groupings (or clusters) of observations (Kaufman and Rousseeuw, 1990). It is an exploratory data-analysis technique. The task here is to group occupations by similar skill characteristics. It was organized in two steps. First, the 288 occupations were grouped according to Ward's linkage procedure (Ward, 1963). From step to step, Ward's linkage joins pairwise clusters with the smallest increase in the variance. Second, applying the k-means procedure corrects for the problem of irreversibility in the first step. This problem is typical for hierarchical and agglomerative procedures as Ward's linkage. While in the course of the procedure the average variable values in the groupings change, it is possible that some cluster objects which already belong to a larger group (cluster) are more similar to other occupation groups than their actual one ('moving centroids'). Since the groups are not split up by Ward's linkage in this case, it is necessary to apply k-means, which corrects the groupings by using the cluster solution of the first step as a starting point. Only 20 occupations were re-ordered by the k-means procedure, which is a rather low rate.

Descriptive Results

The following description of cluster analysis refers to the employment statistics of the Federal Employment Agency, which is also used in the next section. It covers all employees subject to social insurance contribution. The corresponding data set IAB Pallas Online was accessible due to cooperation with the Institute of Employment Research (IAB). Individual data were aggregated to the level of industries and districts keeping information about age and wage, firm size and sex. The five qualification classes (qc) – as the cluster solution is called in the following – had to be transcoded from the occupational classification used in the Microcensus ('Klassifikation der Berufe 1992', Kldb92) to the employment statistics' classification ('Klassifikation der Berufe 1988', KdB88) at the basis of a common four-digit level code.

The distribution of vocational education degrees by qualification class is shown in absolute terms of workers in Figure 11.1. As described above, vocational education entered the analysis in the form of years of schooling

– combined with general education. Since vocational education is the most used skill characteristic, the qualification classes are described by their vocational education structure. The other skill variables are similarly distributed. The skill level is monotonously decreasing in order, qualification classes 4 and 5 are to be characterized as low-skilled occupations in contrast to the other three 'skilled' qualification classes. Although vocational education is more pronounced in qc4, both classes consistently show relatively high shares of employees with lower secondary school level ('*Hauptschulabschluss*'). Also, the participation rate in vocational training almost doubles between each qualification class (not shown in the figure).

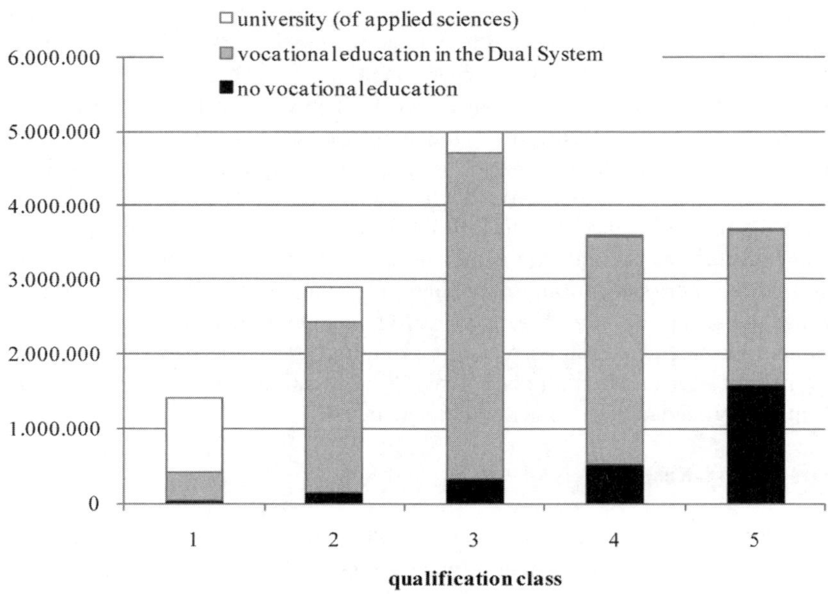

Source: IAB Pallas Online. Author's calculations.

Figure 11.1 Employment and educational structure by qualification class 2006

Employment growth by qualification class from 2000 to 2006 is depicted in Figure 11.2. Note from this figure that employment in low-skilled (qc4, qc5) occupations decreases and skilled occupations including those with high shares of DSVE graduates (q2) instead or university graduates (qc1) remain stable.

The finding that other qualification classes with a relatively high share of DSVE graduates (q3, qc4) face negative employment growth is an important

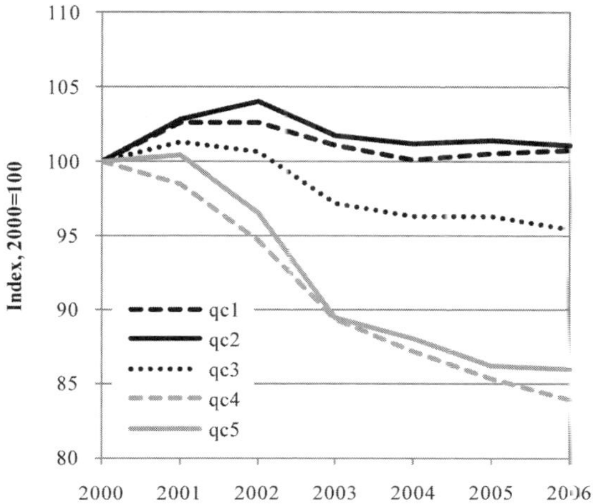

Source: IAB Pallas Online. Author's calculations.

Figure 11.2 Employment growth in five qualification classes 2000–2006

feature of the qualification classes. Thus, using the level of vocational education is misleading in the face of the different development in the middling occupation classes which consist of comparable shares of employees with an intermediate level. This heterogeneity of the intermediate educational level is consistent with the work by Autor et al. (2003), who emphasized that the content of work (the tasks) is more important for possible substitutive effects of technological change than the educational level because tasks of a routine nature are more easily replaced by machines and algorithms. Similarly, these tasks of the intermediate educational level are also more easily dislocated (see, e.g. Grossmann and Rossi-Hansberg, 2008)

SHIFT-SHARE REGRESSION OF QUALIFICATION-SPECIFIC EMPLOYMENT GROWTH

Empirical Model

Using the above-mentioned database IAB Pallas Online, we regress yearly employment growth from 2000 to 2006 in each of the five qualification classes at the level of 12 industries (Table 11.1) within 346 West German districts (NUTS3). By analyzing employment growth at the local industry level the model can be described as a multilevel approach which allows testing

Table 11.1 Industry classification

Industry	German classification of economic activities, ed. 2003
Agriculture, forestry, fishing	011–050
Construction, mining	101–145, 451–455
Low-tech manufacturing	151–160, 171–193, 261–268
Rationalization-intensive low-tech manufacturing	201–223, 251–252, 271–287, 361–366
High-level technology	231–243, 245–247, 291–297, 311–316, 341–355
Leading edge technology	244, 300, 321–335
Personal/household services	501–505, 521–522, 524–527, 551–555, 633, 921–950
Business services	511–519, 651–714, 731–748
Communication services	643, 721–726, 221
Transport/post, energy/water supply	601–641, 371–410, 900
Health services	523, 851–852
Public services	751–804, 853, 911–913, 990

Author's compilation

intra-industrial spatial division of labor via the region type effects. So the subject of analysis is the employment growth of an industry in a region instead of the usually regarded employment growth in the region as a whole.

The five panels, one for each qualification class, consisted of slightly more than 23 000 observations. The explanatory variables of interest will be the region's density and skill-specific concentration indices, but also the regionally fitted values for the industry effect.

Our empirical approach is an econometric analogue to the shift-share decomposition technique as suggested by Patterson (1991) and developed by Möller and Tassinopoulos (2000) and Blien and Haas (2005). See these references for more technical details. In its basic form, employment growth and the industry-district level are regressed on industry, region and year dummies.

The major components of the OLS regression of yearly employment growth in industry-region cells are dummies for each region, industry and year. Other included variables are the local industrial qualification structure in terms of the five qualification classes, the firm size structure within the regional industry captured by three employment classes with class limits of 20 and 100 employees and finally the region type. The region type effect is measured by nine dummy variables emanating from a classification of population density and the higher order region type (e.g. agglomeration or rural areas, see Table 11.2). The regression formula thus consists of almost 400 regressors: 346 district dummies (understood as fixed effects), 12 industry dummies, nine region type dummies, three firm size employment shares, regional industrial employment shares in each of the five qualification classes, and regional wage level in interaction with 12 industry dummies. The regional wage level results from a regression of log regional average wage on the average age, share of employed women, firm size classes, region type dummies and region effects. These region effects from the wage regression serve as wage level without endogeneity problems.

Table 11.2 Definition of region types

Region base type	Region (county) type	Definition
Base type I: **Regions with large agglomerations**	1 Core cities	City districts >100 000 inhabitants (inh.)
	2 Highly urbanized districts	Districts with density ≥ 300 inh/km²
	3 Urbanized districts	Districts with density ≥ 150 inh/km²
	4 Rural districts	Districts/district regions density < 150 inh/km²
Base type II: **Regions with intermediate agglomerations**	5 Central cities	City districts > 100 000 inh.
	6 Urbanized districts	Districts/district regions density ≥ 150 inh/km²
	7 Rural districts	Districts/district regions density < 150 inh/km²
Base type III: **Rural regions**	8 Urbanized districts	Districts/district regions density ≥ 100 inh/km²
	9 Rural districts	Districts/district regions density < 100 inh/km²

Source: Federal Institute for Research on Building, Urban Affairs and Spatial Development (BBSR)

In order to avoid perfect multicollinearity within each component set, restrictions are imposed setting the weighted sums of all concerned variables to zero. The weights are employment shares in the middle of the period, i.e. 2003. All variables are also weighted to avoid heteroscedasticity: since growth rates are fundamentally dependent on their initial level, thus generating heteroscedasticity.

Results of the Baseline Specification

In the following, we refer to five regression analyses of employment growth in each qualification class. Afterwards, we estimate an extended model which searches for skill-specific concentration effects.

Components of Qualification-specific Regional Employment Growth

Results for the baseline specification are shown in Table 11.3. The coefficients report the percentage change of the growth rate in the concerned employment group due to an incremental increase in the variable value.

Following the regression results for skilled occupations (qc1–qc3) we find positive growth effects of high-tech manufacturing, communication, health and public services which amount up to 13.4 percent of yearly employment growth. So there is further growth in industries which already have relatively high shares of skilled labor. Regarding qualification classes 4 and 5, there are only a few cases where the service sectors promote employment, whereas agriculture/forestry/fishing is clearly positively developing (2.4 percent and 4.0 percent, respectively). From the manufacturing sector only the two rather knowledge-intensive industries are strengthening low-skilled employment growth.

Consistent with the theory of functional specialization within industries, population density has different effects on skill-specific employment growth. The effects differ especially between core cities and districts of less density which favor employment growth either for skilled occupations (core cities) or less skilled occupations (districts with fewer inhabitants per km^2). Furthermore, central cities have a consistently positive employment effect. The difference between the two city types is that core cities and central cities (types 1 and 5) both have a population size of at least 100 000 inhabitants, but they differ in the density of their neighboring regions, which is lower for central cities by definition.

Like core cities, highly urbanized districts exhibit negative employment effects for less skilled occupations. In contrast, all other more peripheral and rural region types are beneficial for employment growth in the qualification classes 4 and 5. Following theoretical considerations, the different region type effects probably result from an intra-industry division of labor.

Assessment of Region Type Effects vs. Industry Effects by Region Type

The structure of the model allows quantifying the local industry. In our context it is interesting to compare the (fitted) average effect of the industry mix in a certain region type with the region type effect itself, which expresses the intra-industrial division of labor. The median industry effects summed up with the regional effect are shown in Table 11.4. The latter can be directly gained from the coefficient estimates of the nine region type dummies.

In general, central cities tend to have a more favorable industry mix than other region types, especially compared to core cities within larger agglomerations. The median industry effect in city districts is between –0.05 (qc4) and 0.28 percent yearly employment growth (qc3). Even the negative median industry effect for the less skilled qualification class 4 is only at the third position compared to industry effects in the other region types: for example, the median industry effect in core cities is –0.19 percent. Thus, the median industry mix in core cities in large agglomerations is not a general employment driver. Only for qc3 and qc5 are there positive industry effects on the yearly employment growth of 0.09 and 0.05 percent, respectively.

Adding the region type effect to the median industry effect expresses the composite effect of inter- and intra-industrial division of labor. These results are shown in the second column of each qualification class in Table 11.4 and make up to 2.45 percent employment growth per year. The following pattern appears.

- Core cities are more beneficial for skilled occupations (qc1, qc2, qc3), that is, region type plus industry effect on employment growth is more than average. Less skilled occupations develop worse than average due to the region type and industry mix.
- Central cities promote employment growth for all qualifications.
- In districts of less population density (so-called urbanized and rural districts) employment growth of skilled occupations is less than average but higher for less skilled occupations.

In conclusion, there is a clear skill-related division of labor between core cities in large agglomerations promoting skilled employment growth and districts with lower population density with more favorable employment perspectives for less skilled occupations. The favorable industry and region type effects in central cities concerning occupations from most qualification classes remain the exemption from the rule.

Concentration Processes

The baseline specification is now extended by measures of industrial concentration processes within each qualification class. The variable of interest is constructed as the difference between total and local proportions of qualification-specific employment in the concerned industry in the initial year 2000 (see Möller and Tassinopoulos, 2000). It measures the impact of qualification-specific

Table 11.3 Shift-share regression estimation results

Independent variable	qc1 (high-skilled)	qc2	qc3	qc4	qc5 (low-skilled)
Core cities (I)	0.005 ***	0.005 ***	0.002 **	–0.006 ***	–0.011 ***
Highly urbanized districts (I)	–0.001	0.000	0.000	–0.006 ***	–0.005 ***
Urbanized districts (I)	–0.003	–0.006 **	–0.002	0.000	0.005 **
Rural districts (I)	–0.013	–0.007	–0.005	0.003	0.001
Central cities (II)	0.007 ***	0.001	0.004 ***	0.009 ***	0.009 ***
Urbanized district (II)	–0.007 ***	–0.003 *	–0.002 **	0.001	0.004 **
Rural districts (II)	–0.013 ***	–0.007 ***	–0.004 **	0.007 **	0.009 ***
Urbanized districts (III)	–0.009 **	–0.007 ***	–0.002	0.010 ***	0.007 ***
Rural districts (III)	–0.020 ***	–0.008 **	–0.008 ***	0.006	0.009 **
Agriculture, forestry, fishing	–0.015	–0.025	–0.026 **	0.024 ***	0.040 ***
Construction, mining	–0.058 ***	–0.038 ***	–0.050 ***	–0.016 ***	–0.058 ***
Low-tech manufacturing	–0.052 ***	–0.040 ***	–0.046 ***	0.008 *	–0.012 ***
Rationalization-intensive low-tech Manufacturing	–0.033 ***	–0.019 ***	–0.039 ***	0.000	0.003

High-level technology	0.024 ***	0.015 ***	0.001	0.028 ***	0.011 ***
Leading edge technology	0.022 ***	0.005	0.006 **	0.034 ***	0.004
Personal/household services	–0.057 ***	–0.028 ***	–0.021 ***	–0.012 ***	–0.013 ***
Business services	–0.013 ***	–0.012 ***	0.005 ***	–0.005	0.013 ***
Communication services	0.028 ***	0.023 ***	0.030 ***	0.132 ***	–0.020
Transport/post, energy/water supply	–0.025 ***	–0.032 ***	–0.019 ***	–0.015 **	0.033 ***
Health services	0.018 ***	0.011 ***	0.018 ***	0.009	–0.044 ***
Public services	0.010 ***	0.013 ***	0.025 ***	–0.018 **	–0.029 ***

Notes:
Excerpt from the regression output
Not reported: industrial employment shares of the qualification classes, industry-wage interaction terms, firm size employment shares, region fixed effects, year dummies.
***/**/* indicates significance at the 1%/5%/10% level
See Table 11.2 for the definition of region types

Source: IAB Pallas Online. Author's calculations

Table 11.4 Median industry effects by region type

Region type (base type)	qc1		qc2		qc3		qc4		qc5	
	Industry effect	Industry effect Region type effect	Industry effect	Industry effect + Region type effect	Industry effect	Industry effect + Region type effect	Industry effect	Industry effect + Region type effect	Industry effect	Industry effect + Region type effect
Core cities (I)	–0.09	0.45	–0.01	0.51	0.09	0.32	–0.19	–0.77	0.05	–1.03
Highly urbanized districts (I)	–0.22	–0.32	0.05	0.02	–0.15	–0.12	0.02	–0.55	0.11	–0.38
Urbanized districts (I)	–0.32	–0.66	0.08	–0.52	–0.14	–0.32	–0.18	–0.17	–0.13	0.39
Rural districts (I)	–0.36	–1.66	0.23	–0.51	–0.04	–0.58	–0.41	–0.16	–0.21	–0.06
Central cities (II)	0.08	0.81	0.11	0.19	0.28	0.71	–0.05	0.87	0.05	0.97
Urbanized districts (II)	–0.07	–0.77	0.12	–0.19	–0.20	–0.43	0.00	0.08	–0.16	0.21
Rural districts (II)	–0.33	–1.60	0.10	–0.62	–0.37	–0.72	–0.08	0.62	–0.36	0.54
Urbanized districts (III)	–0.27	–1.14	0.07	–0.63	–0.23	–0.39	–0.10	0.86	–0.36	0.31
Rural districts (III)	–0.43	–2.45	0.08	–0.76	–0.33	–1.08	–0.22	0.33	–0.43	0.49

Source: IAB Pallas Online. Author's calculations

See Table 11.2 for the definition of region types

employment specialization in the concerning industry. Positive coefficients indicate accelerated growth where the concerned qualification class is more strongly represented. Negative coefficient estimates thus indicate upward adjustments of employment where the occupations in the concerned industry are underrepresented and downward adjustments in case of relative specialization, respectively. From the estimates we are then able to derive first, whether specialization is beneficial (e.g. due to knowledge spillover) for all qualification classes or only the skilled ones, and second, whether industrial concentration is behind the aforementioned positive periphery effect on low-skilled employment.

In fact, there are positive coefficient estimates – presumably localization effects – for all qualifications in manufacturing (except for the high-level technology) but most significantly for rationalization-intensive low-tech manufacturing (Table 11.5). In contrast, high-level technology industries are less dependent on concentration: the coefficients either show no clear skill-specific pattern or are not significant. So attracting high-technology manufacturing is beneficial for employment as the industry effects of the baseline specification show, but it is not necessary to initiate special cluster oriented measures for these industries since there is no additional pay-off from concentration.

Moreover, high- and mid-skilled occupations are increasingly concentrated in other industries, even in service industries where cluster policies usually are less popular. These concentration forces in skilled occupations can be explained by positive externalities within the same qualification classes such as pooled and specialized labor markets or knowledge spillover effects. In contrast, low-skilled in services tends to disperse spatially: This is the case for qc4 occupations and also for qc5 occupations in business services or communication services. However, this negative sign indicates catching-up processes of the local industry mix as well as downsizing processes of concentrated locations.

Concerning region type effects, the coefficient's signs are not very different from the preceding regression analysis. Thus, there is no hidden concentration effect for specific industries in the preceding regression results for the region type effect. The segmentation pattern remains stable even when controlling for industry and qualification-specific concentration processes. The theory of functional specialization which predicts a qualification-related spatial division of labor due to increasing concentration within qualification classes is therefore only partially supported since we still find the division of labor pattern in this extended model.

On the one hand, we find concentration effects for skilled occupations. On the other hand, lower skilled employment decreases faster if initial employment shares are higher. Conversely, employment increases where employment shares are less than average. Thus, for less skilled occupations there are negative concentration effects. While spatial distribution of skilled employment is increasingly concentrated, less skilled employment tends to distribute more evenly. The fact that urban areas only favor employment growth in skilled

Table 11.5 Coefficient estimates of industry-specific (de-)concentration processes

Independent variable	qc1		qc2		qc3		qc4		qc5	
Agriculture, forestry, fishing (conc.)	0.781	*	1.750	**	1.268	**	0.171		–1.100	
Construction, mining (conc.)	0.316	**	0.257		0.300	**	0.050		0.101	
Low-tech manufacturing (conc.)	0.317	*	0.753		0.239	***	0.030		0.029	
Rationalization-intensive low-tech manufacturing (conc.)	0.195	***	0.425	***	0.169	***	0.126	***	0.075	***
High-level technology (conc.)	0.024		–0.070	***	0.061	***	–0.056	**	–0.015	
Leading edge technology (conc.)	0.017		0.425	***	0.126		0.069		–0.228	**
Personal/household services (conc.)	0.067		0.532	***	0.072		0.066		0.131	***
Business services (conc.)	0.001		0.147	***	0.031	**	–0.142	**	–0.091	***
Communication services (conc.)	0.134		0.392	***	–0.421	**	–0.215		–4.347	***
Transport/post, energy/water supply (conc.)	0.359	***	0.276	***	0.350	***	–0.569	*	0.102	***
Health services (conc.)	0.107	*	0.163	***	0.007		–0.937		0.264	
Public services (conc.)	0.036		0.079	***	0.009		0.011		0.032	
Core cities (I)	0.010	***	0.012	***	0.006	***	–0.004	*	–0.008	***
Highly urbanized districts (I)	–0.001		–0.004	***	0.000		–0.006	***	–0.006	***
Urbanized districts (I)	–0.007	*	–0.013	***	–0.004	**	–0.001		0.004	
Rural districts (I)	–0.019	**	–0.017	***	–0.009	**	0.001		0.000	
Central cities (II)	0.008	***	0.008	***	0.005	***	0.012	***	0.013	***
Urbanized districts (II)	–0.012	***	–0.009	***	–0.004	***	0.000		0.003	*

Rural districts (II)	–0.021	***	–0.016	***	–0.008	***	0.006	*	0.007	***
Urbanized districts (III)	–0.016	***	–0.014	***	–0.005	***	0.009	***	0.006	**
Rural districts (III)	–0.032	***	–0.017	***	–0.014	***	0.005		0.007	*
Agriculture, forestry, fishing	–0.055	***	–0.094	***	–0.065	***	0.021	**	0.040	***
Construction, mining	–0.084	***	–0.091	***	–0.068	***	–0.018	***	–0.072	***
Low-tech manufacturing	–0.068	***	–0.066	***	–0.059	***	0.010	**	–0.014	***
Rationalization-intensive low-tech manufacturing	–0.045	***	–0.045	***	–0.048	***	–0.001		–0.001	
High-level technology	0.027	***	0.020	***	0.000		0.033	***	0.015	***
Leading edge technology	0.028	***	0.004		0.007	**	0.035	***	0.020	***
Personal/household services	–0.072	***	–0.077	***	–0.028	***	–0.016	***	–0.024	***
Business services	–0.014	***	–0.015	***	0.007	***	–0.004		0.019	***
Communication services	0.029	***	0.048	***	0.038	***	0.134	***	0.030	*
Transport/post, energy/water supply	–0.036	***	–0.057	***	–0.027	***	–0.008		0.034	***
Health services	0.021	***	0.022	***	0.026	***	0.009		–0.037	***
Public services	0.018	***	0.020	***	0.033	***	–0.017	**	–0.021	***

Notes:
Excerpt from the regression output
Not reported: industrial employment shares of the qualification classes, industry-wage interaction terms, firm size employment shares, region fixed effects, year dummies
***/**/* indicates significance at the 1%/5%/10% level
See Table 11.2 for the definition of region types
'conc.' stands for the concentration effect in the related industry

Source: IAB Pallas Online. Author's calculations

occupations, in contrast to rural areas with favorable conditions for the growth of low-skilled employment, remains unchanged. To conclude, industry-specific concentration of different qualifications is not bound to certain region types. Therefore, we are facing two phenomena of spatial segmentation: one that is related to the region type and another that is related to the respective industry.

SUMMARY AND DISCUSSION

Employment growth in less skilled occupations has got little focus in the empirical literature although high unemployment risks and low regional mobility of these groups often constitute a social burden for a region's underlying strong structural changes. This is even more surprising as we consider the employment–productivity contradiction at the local level (Cingano and Schivardi, 2004; Almeida, 2007) which could be explained by skill-biased regional employment growth. However, from a theoretical point of view, there is the possibility of positive localization effects on low-skilled employment due to a (skill-related) regional 'functional specialization' (Duranton and Puga, 2005). In this chapter we try to assess the regional determinants of qualification-specific employment growth against this background.

The five qualification classes and their yearly employment growth at the level of industries and districts between the years 2000 and 2006 are subject to a regression analogue of the common shift-share technique. The most interesting finding is a division of labor: cities promote skilled employment growth whereas less skilled occupations are better off in peripheral areas. This pattern also holds if concentration effects of skill-specific employment are controlled for. Concentration effects are only found for skilled occupations while occupations with minor or without requirements of formal vocational education show negative concentration effects and therefore tend to disperse spatially.

One interesting exception is the positive concentration effect for the two less-skilled occupation groups in the rationalization-intensive low-tech industries which are defined by high rates of cost-reducing innovation (Rammer, 2006). It can be concluded that the rationalization efforts in this sector also exploit localization advantages like functional specialization.

In summary, there are two main empirical observations regarding skill-specific regional employment growth. First, employment in skilled occupations grows in cities more than average while – vice versa – less skilled occupations benefit develop better in rural areas. This observation can be described as spatial division of human capital or segmentation which is not resulting from the industry mix, but predominantly from intra-industrial division of labor. Second, only skilled occupations are spatially concentrated within industries. This argument maintains that localization effects such as knowledge spillovers or labor market pooling are not generally found for less skilled occupations. Since the region type effect (which indicates the intra-industrial division of labor) remains unchanged if controlled for the industry-specific concentration, both effects are two separate empirical observations. These results contradict the expectations derived from the theory of functional specialization, which states that localization effects for less skilled manual labor exist beside the cost advantages arising from a peripheral location.

The policy conclusions are twofold. First, development strategies focusing on human capital promotion or cluster strategies neglect that employment opportunities for less skilled workers are then decreasing. For example, localized endogenous skill-biased technological change (Berry and Glaese, 2005; Beaudry et al., 2006; Almeida, 2007) provides one explanation for the unfavorable employment conditions of less skilled occupations in cities. Yet, it is unclear whether the positive periphery effect on less skilled employment growth simply results from the strong negative city effect and that the structural change in rural areas is just slower or whether there are real employment prospects with absolute growth. Against this background it will be interesting to observe if the peripheral regions in developed countries are going to benefit from the relocation out of the 'international periphery' represented by low-cost countries (see, e.g., Goel et al., 2008). Similarly to Massey (1994, p. 47), who once concluded 'that [a contradiction] is not between inner cities and policy-aided development areas, but between the cities and the demands of profitability and international competitiveness' our second conclusion here would be to consider a competition between peripheral areas in developing and developed countries. Since the main regional determinants driving the knowledge economy (urban context, location of knowledge-intensive sectors, industrial clusters) rarely exhibit positive employment effects for low-skilled workers it is more important to exploit other locational advantages, especially in the periphery, in order to avoid further unemployment.

REFERENCES

Aarland, K., J.C. Davis, J.V. Henderson and Y. Ono (2007), 'Spatial organization of firms: the decision to split production and administration', *RAND Journal of Economics*, **38**, 480–94.

Almeida, R. (2007), 'Local economic structure and growth', *Spatial Economic Analysis*, **2**, 65–90.

Amiti, M. and C.A. Pissarides (2005), 'Trade and industrial location with heterogeneous labor', *Journal of International Economics*, **67**, 392–412.

Autor, D.H., F. Levy and R.J. Murnane (2003), 'The skill content of recent technological change: an empirical exploration', *The Quarterly Journal of Economics*, **118**, 1279–1333.

Beaudry, P., M. Doms and E. Lewis (2006), *Endogenous Skill Bias in Technology Adoption: City-Level Evidence from the IT Revolution*, NBER Working Paper No. 12 521.

Berry, C.R. and E.L. Glaeser (2005), 'The divergence of human capital levels across cities', *Papers in Regional Science*, **84**, 407–44.
Blien, U. and A. Haas (2005), 'Service industries and regional development: an analysis for eastern Germany', *Service Industries Journal*, **25**, 979–97.
Cingano, F. and F. Schivardi (2004), 'Identifying the sources of local productivity growth', *Journal of the European Economic Association*, **2**, 720–42.
Defever, F. (2006), 'Functional fragmentation and the location of multinational firms in the enlarged Europe', *Regional Science and Urban Economics*, **36**, 658–77.
Duranton, G. and D. Puga (2001), 'Nursery cities: urban diversity, process innovation, and the life cycle of products', *American Economic Review*, **91**, 1454–77.
Duranton, G. and D. Puga (2005), 'From sectoral to functional urban specialization', *Journal of Urban Economics*, **57**, 343–70.
Fujita, M. and J-F. Thisse (2002), *Economics of Agglomeration*, Cambridge: Cambridge University Press.
Goel, A., N. Moussavi and V.N. Srivatsan (2008), 'Time to rethink offshoring?', *The McKinsey Quarterly*, September 2008.
Grossman, G.M. and E. Rossi-Hansberg (2008), 'Trading tasks: a simple theory of offshoring', *American Economic Review*, **98**, 1978–97.
Kaufman, L. and P.J. Rousseeuw (1990), *Finding Groups in Data*, New York: Wiley.
Leamer, E.E. and M. Storper (2001), 'The economic geography of the internet age', *Journal of International Business Studies*, **32**, 641–65.
Lovely, M.E., S.S. Rosenthal and S. Sharma (2005), 'Information, agglomeration, and the headquarters of US exporters', *Regional Science and Urban Economics*, **35**, 167–92.
Massey, D. (1994), *Space, Place and Gender*, Minneapolis: University of Minnesota Press.
McCann, P. and J. Simonen (2005), 'Innovation, knowledge spillovers and local labor markets', *Papers in Regional Science*, **84**, 465–85.
Möller, J. and A. Tassinopoulos (2000), 'Zunehmende spezialisierung oder strukturkonvergenz? Eine analyse der sektoralen beschäftigungsentwicklung auf regionaler Ebene', *Jahrbuch für Regionalwissenschaft*, **20**, 1–38.
Mori, T. and A. Turrini (2005), 'Skills, agglomeration and segmentation', *European Economic Review*, **49**, 201–26.
Ota, M. and M. Fujita (1993), 'Communication technologies and spatial organization of multi-unit firms in metropolitan areas', *Regional Science and Urban Economics*, **23**, 695–729.
Patterson, M.G. (1991), 'A note on the formulation of a full-analog regression-model of the shift-share method', *Journal of Regional Science*, **31**, 211–16.
Power, D. and M. Lundmark (2004), 'Working through knowledge pools: labor market dynamics, the transference of knowledge and ideas, and industrial clusters', *Urban Studies*, **41**, 1025–44.
Rammer, C. (2006), 'Innovation in firms', in U. Schmoch, C. Rammer and H. Legler (eds), *National Systems of Innovation in Comparison. Structure and Performance Indicators for Knowledge Societies*, Dordrecht: Springer, pp. 107–32.
Vernon, R. (1966), 'International investment and international trade in the product cycle', *Quarterly Journal of Economics*, **80**, 190–207.
Ward, J.H. Jr (1963), 'Hierarchical grouping to optimize an objective function', *Journal of the American Statistical Association*, **58**, 236–44.

PART IV

Clusters and productivity

12. Clusters and cluster development in the knowledge economy

Andreas P. Cornett

INTRODUCTION

Historically, the roots of clusters as a business strategy date back to classical Marshallian and Italian concepts of industrial districts. Traditionally, the concepts of industrial districts and clustering have been limited to the manufacturing sector. Today, the concept is applied in a much broader context, covering all kinds of economic activities, including private and public services with relevance for economic development and performance. Furthermore, the concepts have been extended to cover a wide range of activities related to economic policy and the improvement of regional competitiveness and growth (i.e. knowledge and innovation transfer and dissemination as well as the support of research and development). Triple helix inspired frameworks are often central to cluster-oriented development policy. They provide a mechanism to combine three main elements: innovation, entrepreneurship and clustering. To be sure, the concept of clustering has become increasingly popular in business development policy. The combination of these aspects leads to the central research question of this chapter, to assess the role and potential of a cluster development policy in a knowledge economy.

Cluster-oriented development policy is the focus of this chapter, which is organized into four subsequent sections. First, traditional concepts of clustering are presented. Next, more recent approaches incorporating knowledge-based components are discussed using a conceptualization of 'space-less' or virtual processes focused on how regional cluster-based development policy can be linked to the global economic and knowledge system. Today, this is crucial in a world of new international division of production and functional division of labor. The next section introduces current Danish regional cluster policies and analyses the role of these policies on regional development in Southern Denmark. The final section provides a summary of the analytical results, relevant policy implications, and future conceptual and empirical research needs.

CONCEPTS OF CLUSTERING

Clusters or industrial districts are well-established concepts in economic geography and regional science. In the last two decades, the concepts have experienced a rebirth from a policy perspective focusing on the ability to improve economic performance and regional competitiveness. After a brief literature review of cluster theory, this section extends a framework that combines these concepts with knowledge and competitiveness. The aim is to assess the theoretical arguments that support a belief among business development policy makers that clustered industries perform better than firms located outside the cluster area.

Originally, the concept of industrial districts was related to the way manufacturing industries were organized, as first described by Marshall in the late nineteenth century who developed the concept of industrial districts based on location advantages and what currently is known as positive pecuniary externalities based on the skills of a common labor pool (Marshall, 1947, p. 267).

The Italian industrial districts in the Emilia-Romagna in particular are well-known and have been the target of many in-depth analyses, but also other types of clusters (i.e. high-tech clusters) have been studied, like the Silicon Valley or the Cambridge area in the UK. In a study of Italian industrial districts, Signorini stresses the important features of industrial districts as follows (Signorini 1994, p. 370):

- the division of labor, through inter-firm specialization, improves productivity and flexibility;
- the environmental or milieu aspect, dealing with cultural factors (attitudes, knowledge and behavior) and infrastructure (physical as well as the availability of service);
- the network element, that is, forward and backward linkages reinforcing the potential for the district as a whole.

Particularly the second and to some extent the third element are relevant targets of a business development policy to improve competitiveness and to overcome regional disadvantages by reinforcement of the interrelatedness of business and industry in a specific region. A consequence of this broader perspective is that the concept of industrial districts has to be elaborated in three ways. First, the concepts are enlarged in scope in order to deal with sectors other than manufacturing and production. A modern system of production and innovation goes beyond the traditional division of labor within an industry or branch: business service is also of increasing importance for the regional economic base (Illeris, 1999). Second, a distinction between industrial clusters (agglomeration of industries or sites of production) and 'true' industrial districts[1] or modern types of 'virtual' or non-spatial cluster processes are developed due to the fact that present production systems are not necessarily based in an adjacent geography. The third element is to focus on non-firm actors as part of the cluster landscape like knowledge institutions and the public sector as supportive measures in a proactive business development policy.

Central elements in the application of the cluster concept in a development and policy context include:

- the characteristics of clusters and traditional industrial districts;
- the nature of systems of production with regard to regional and/or functional characteristics;
- the organizational set-up of a system as a supportive measure for cluster policy.

The latter is particularly important in the broader concept of clustering used in recent studies of the performance of clustered industries and regions focusing on the presence of clusters or geographic concentration of industries linked together. According to Porter (2003, p. 562), a cluster is defined:

> as a geographically proximate group of interconnected companies, suppliers, service providers and associated institutions in a particular field, linked by externalities of various types … Clusters are important because of the externalities that connect the constituent industries, such as common technologies, skills, knowledge and purchased inputs.

It is important to note that Porter explicitly stressed that an industry can become a part of several clusters, which can cause problems in the empirical assessment of cluster performance (i.e. with regard to employment). From a conceptual point of view the issue is solved by a distinction between narrow and broad cluster definition. The former is based on the strongest geographical and location ties of a particular industry. Broad clusters include all industries in the cluster (Porter, 2003, p. 563), see also Porter (1998, p.199) for a general elaboration of the concept of clusters. In this sense, the cluster shares many features with the more general concepts of production systems (or from a firm level, value chains). This narrow concept shares many features with mainstream industrial district literature, focusing very much on the inter-firm networks in a specific sector or area. For a useful summary of the concepts, the theoretical foundation and an assessment of the sources of economic impacts of clustered regional systems of production, see Bergman and Feser (1999). Traditionally, industrial districts are defined sectorally, geographically and functionally. All three criteria have to be fulfilled to a certain degree. With the ongoing internationalization of the value chain (i.e. outsourcing and relocation of production sites) the spatial rigidity of the cluster concept seems to disappear in many sectors or industries. It can be questioned whether the geographical precondition is necessary, or whether it is possible to take advantage of the potentials of positive pecuniary externalities ascribed to clusters or industrial districts in a spaceless or virtual context. In the age of globalization and outsourcing, the production chain in clustered industries has also changed with significant and often dominant production taking place outside the geographical cluster. However, firms in the clusters often remain at the core of the value chain and become engines of growth and industrial renewal.

The definition quoted below is only partly relevant in the current context, but still stresses the importance of industrial districts not being limited to traditional manufacturing activities. Park and Markusen defined an industrial district as follows: 'an industrial district is a sizable and spatially delimited area of trade-oriented economic activity which has a distinctive economic specialization, be it resource-related, manufacturing, or services' (Markusen, 1996, p. 296). The important message of the quoted definition is that the concept goes beyond traditional manufacturing-based approaches. Another alteration of the traditional concept introduced in this chapter is the inclusion of non-firm actors, specifically when the industrial districts or clusters are analysed as innovative and knowledge systems. The latter seems to be implicit in the aims of newer concepts on business policy frameworks introduced by proactive governmental agencies in Denmark in the last decade, see Ministry of Industry (1998, pp. 198–266) and Erhvervs og Boligstyrelsen (2003).

Pickernell et al. (2007) have further elaborated on the concept of clusters and networks for economic development purposes. Based on a survey of cluster and network literature, eight types of cluster frameworks are identified, ranging from the traditional concepts discussed above to non-spatial networks and clustering processes (see Table 1 in Pickernell et al., 2007, p. 243). This approach provides an ability to include a broader scope of measures and instruments partly grounded in new economic growth theory. The classification developed in Pickernell et al. (2007) provides a useful starting point to link traditional cluster concepts to the drivers of economic growth and in particular the integration of human capital, innovation and the knowledge economy.

In this perspective, the role of knowledge and innovation is at least twofold. In the short run, innovation and knowledge creation become necessary instruments in the process of regional adaptation to industrial change as a response to changes in the competitive environment. One example of the new focus on competence is the ongoing discussion of knowledge as a central factor in the spatial organization of industries (Malmberg and Maskell, 2002, p. 438).

The focal point of the concept is the interaction between the business entities and the external environment as a part of a broader network of innovative relations covering intra-firm as well as extra-firm relations and processes. Regardless, the fact that clusters are usually only a small part of the regional economy, and policy has to be seen within the context of the general innovation and knowledge policy, clusters and cluster policy are often considered to be an efficient tool in transferring policy into economic growth and competitiveness. For this reason clustering instruments play an increasingly prominent role in business development and industrial policy:

> Second, this task has obvious policy relevance today. Throughout the OECD world (and beyond, as a matter of fact) cluster-based policies have in recent years increasingly been seen as the main option in the field of industrial and regional policy. As an important element of these policies we find a doctrine saying that regions should specialize industrially and promote the dynamics of spatial clustering in order to gain or sustain competitiveness and prosperity. (Malmberg and Maskell, 2002, p. 431)

In the context of this chapter the appropriate analytical point of departure is an instrumental approach combining all above-mentioned aspects in the elaboration of the cluster concept. According to the instrumental approach, sketched in Pickernell et al. (2007), the most relevant types are 'virtual organizations' (7) and 'social networks' (6), both knowledge-based and characterized by horizontal relations within and across industries (ibid., p. 344). In particular, the 'virtual organization' type seems to be important as a supplement to traditional cluster-policy, enabling the inclusion of non-industry actors (e.g. in a 'triple helix' framework, see Figure 12.1 on p. 219). Type 6 mentioned above, the social network, is delimited as follows:

> Celle 6 [Social networks] is indicative of the processes required for informal horizontal cluster and network types, such as social networks of entrepreneurs with lower power dependence, as exist in some cluster associations. The strategic focus is based around knowledge sharing and dissemination, as would be expected where small firms cooperate to generate scale effects. (Pickernell et al., 2007, p. 350)

These two types of cluster concepts seem to be useful as conceptual anchors for the analysis because they can serve as a bridge between traditional cluster terminology and the broader 'triple helix' framework presented in the next section.

The remainder of this chapter is dedicated to the central research question of how to specify non-spatia-based clusters and the issue of how to transfer the concept into an instrument for a practical-oriented business development policy. The focus is to identify drivers of knowledge and innovation-based activities and how these can be transmitted into regional economic growth and development.

Nevertheless, it has to be stressed that the regional system has to be seen as a whole, including the other drivers and linkages. Clusters and local innovation systems are integral parts of the economic base concept in a new order, dealing with innovation, regional growth and in particular the development of a regional knowledge and learning economy (see also Cheshire and Malecki,

2004, pp. 256–60). In this context the analytical set-up is familiar to the cluster category 'social-network model' discussed in Gordon and McCann (2000, pp. 519–21) and Malmberg and Maskell's concept of knowledge-based clusters (Malmberg and Maskell, 2002, pp. 438–42).

CLUSTER DEVELOPMENT IN A KNOWLEDGE ECONOMY: FROM REGIONAL TO VIRTUAL SPACE?

Traditional clusters are grounded in space, and the geographic location delineates the cluster itself. The geographic aspects of production systems with their foundation in non-material production become increasingly obsolete. Furthermore, material-based production systems are often networks rather than traditional geographical clusters. From a knowledge perspective the non-spatial aspects of newer concepts of clustering are even more obvious. They cannot be exclusively based on traditional physical proximity. For at least two reasons the latter is of major importance when cluster-promotion policy is used as a means of (regional) business development policy. First, while the economic benefits may not always be measurable as regional generated income and employment, this type of integration in an external production and knowledge system, nevertheless, is often a necessary condition for the survival of a local cluster or industrial district in the age of globalization. In particular, increased outsourcing of traditional low-tech branches will become the major challenge in the future. Second, knowledge and technology-intensive industries have reinforced the virtual aspects of traditional clusters or industrial districts. This has changed the conditions for business development in general and the nature of cluster stimulating activities in particular.

The central mechanisms in this process are conceptualized within the framework of the 'triple-helix model' outlined in Figure 12.1. Note from this figure that the 'triple-helix' actors influence the shape and performance of the drivers; in particular entrepreneurship, innovation, human capital and information and communication technology (see Cornett, 2009), and are themselves influenced by the development of the other drivers. Overall, it has to be stressed that the figure represents only a snapshot of conditions at a certain point in time. In reality there exists a dynamic process where drivers and actors interact within a continual sequence of growth from period to period.

A constitutional element in contemporary national and regional business development strategies is the use of clustering as a means to improve and strengthen knowledge creation, dissemination, innovation, and local entrepreneurship. This analysis provides a closer look into the role of cluster-oriented regional and business development policies and attempts to relate the policies to the concepts of clusters presented earlier.

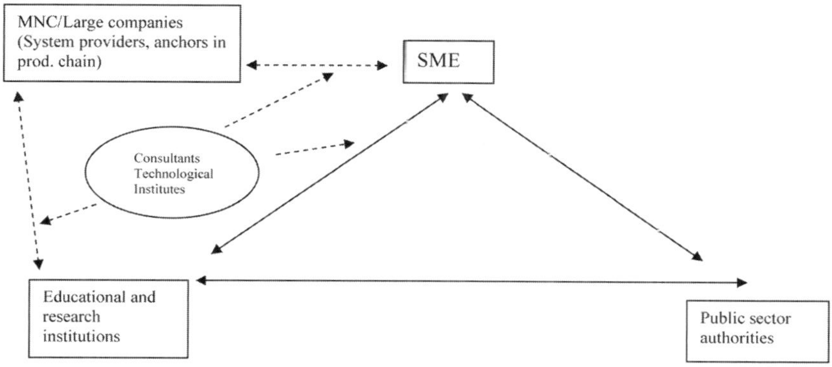

Based on Etzkowitz and Leydesdorff, 2000

Figure 12.1 Partnership of innovation triangle

A central element in this type of cluster policy is the network concept and the provision of an appropriate institutional (not in a material sense) framework for cooperation. In Figure 12.1, note the traditional triad of firms, public sector authorities and the knowledge sector. Analytically, the conceptual point of departure from a traditional 'triple helix' set-up includes the role of different types of companies in the concept. Important units in a proactive cluster policy are enterprises, educational and research institutions and other public authorities. Consultants and advisory organizations are often included in a partnership as well. A 'partnership of innovation' is defined as a functional network aimed at facilitating innovative activity and disseminating the results. This contributes to the development of a competent base with formal as well as tacit knowledge. The latter is often considered the backbone of traditional concepts of clusters or industrial districts. Partnerships of innovation are not bound to a specific spatial context. They are targeted organizational set-ups dedicated to specific objectives (e.g. product or process innovation, regional knowledge management in a broader sense, or the dissemination of new processes or technology within a particular region). Each region or partnership should be able to relate to their specific set-up, highlighting the most appropriate means of communication in the particular region or sector.

In a knowledge economy perspective the partnership concept may be further extended. The principal focus is on the relationship between knowledge institutions and the company sector, often forming a new triangle with the larger companies acting as anchors not only in the production system, but also as knowledge and technology disseminators to subcontractors and small and medium-sized enterprises (SMEs). In this process the advisory system and technological information agencies can be involved as mediators or supportive entities (note the left part of Figure 12.1). From an organizational point of view we have to distinguish between public sector and private sector institutions. This is found in the lower right corner of the triple helix (also, note Appendix 12.1 for an outline of the current Danish system). The baseline of the figure represents a distinction between the applied R&D done by research and higher education institutions and that accomplished by the general public sector.

The partnership triangle outlined in Figure 12.1 has to be analysed in an inter-temporal as well as a spatial and functional context. From a development perspective the 'triple helix' becomes dynamic. The model incorporates impacts of the previous period of development within the system into account (see Etzkowitz and Leydesdorff, 2000, p. 113). The relevant linkages in process development are not necessarily bounded by geography. A cluster-oriented business development policy should be broader than geography. Recent changes of the regional policy targets and measures indicate a change toward a functional orientation as well (see Pickernell et al., 2007, p. 341). In particular, business service agencies or innovation mediators have to serve as local knowledge anchors to transmit new technology and procedures into the local business community. With regard to the educational and knowledge sector, supporting funding for the community service obligations may be an appropriate tool to encourage and improve the transfer of tacit as well as explicit knowledge. In this perspective it seems relevant to quote a phrase used by Steiner (2002, p. 19): 'we need the support of clusters by policy', but at least in Denmark the record has been mixed (Andersen et al., 2006). This represents a challenge met at least partially by the Danish government in the recent past (Ministry of Industry, 2001; Cornett, 2009). The current regional cluster initiatives are discussed in more detail in the next section of this chapter.

A core element of the *regional policy aspect* is the role of clustering as a kind of catalyst in a proactive regional policy. Clusters are often considered to be an efficient tool in transferring policy into economic growth and competitiveness. For a discussion of a regional application of new growth models based on non-linear concepts, see Cornett (2009). From this perspective, the partnership between private and public sectors is of crucial importance especially in areas dominated by small and medium-sized enterprises. Clustering can be initiated by existing firms or through entrepreneurial activities in independent start-ups, as entrepreneurship within the framework of an existing company (intrapreneurship), or initiated by the business development system. The next section highlights this topic based on the current Danish regional development initiative aimed at using cluster development as a means to improve regional competitiveness and performance.

CLUSTER DEVELOPMENT POLICY IN PRACTICE: A DANISH EXAMPLE

During the recent past, economic disparities between European countries have diminished, but at the same time the regional disparities within most countries have increased. The main reason is high growth in central regions and low growth in rural and peripheral areas. This growth difference can also be found in Denmark, not only between the metropolitan region of Copenhagen and the remaining part of the country, but also within the five new regions (Cornett and Sørensen, 2007b). One consequence of this process is that the center–periphery issue has become a central topic in the discussion of the newly launched regional development plans in most Danish regions, and has significant influence on the regional business development policy and the implementation taking place in cooperation with municipalities and other actors.

The regional development strategies in new Danish regions are in many respects regional applications of the general policy set-up provided by the Danish Ministry of Economics and Business, in particular the Regional Growth report launched in May, 2006 (Ministry of Economics and Business, 2006) which in many aspects was based on ideas from new economic growth theories and in particular OECD studies, i.e. OECD (2001). In the process of adopting the overall growth policy strategy to a particular regional setting, the linkages between businesses, the knowledge sector and the public sector is of crucial importance both from an institutional perspective and as a possible dynamic force in restructuring and adaptation processes. For a discussion and specification of this approach in a regional growth and competitiveness context, that is, the importance of entrepreneurship, innovation, human capital and information and communication technology, see Cornett (2009).

Although the regions will face different challenges in future (e.g. compare Western Denmark and Eastern Denmark) the main elements in the regional business development strategies contains many similarities (see Cornett and Sørensen, 2007a). These include with regard to the knowledge economy in particular competency development and cluster related initiatives. Table 12.1 provides an overview of the main targets of the Danish regions' business development strategies.

On the regional level, growth policies have been adapted to the specific regional context and most of the general regional development aims can be found in the regional business development programs. The regional adaptation is noticeably visible in the specific initiatives for the involvement of the knowledge sector (i.e. in the cluster development initiatives). In spite of differences in the level of specification, most of the regions aim to improve or to use entrepreneurship, innovation and clusters in their business development strategy. The most popular instrument is the development of human resources

Table 12.1 *Core elements of the business development strategies of the Danish regions*

Region	Strategies and instruments
Zeeland	Innovation Entrepreneurship Knowledge, education and competences
Copenhagen	Strong competence clusters with rich perspectives New companies – new knowledge Development of the metropolitan region Research and innovation Increase and development of the labor force on all levels Marketing, events, culture and values The Oeresund region in an international perspective
Northern Jutland	Active knowledge institutions Regional innovations milieu Growth entrepreneurs Strong clusters 'Event economy' Sustainable energy and new type of energy in Northern Jutland Outstanding competence development
Central Jutland	Human resources Innovation Entrepreneurship
Southern Denmark	Cluster development 'Event economy' The healthy life Human resources Research, innovation and new technologies Entrepreneurship

Source: Elaboration based on the five regions' business development plans. For details see Vækstforum for Region Hovedstaden (2007), Vækstforum Midt (2007), Vækstforum Region Nordjylland (2007), Vækstforum Sjælland (2007) and Vækstforum Region Syddanmark (2007).

and knowledge institutions. Digging deeper, the differences between targets and levels of ambition vary according to the specific regional environment and the local strongholds that exist within the analysed regions. This is especially the case in the short and medium term. Furthermore, there seems to be a situation of interregional competition between the regions, both in an East–West dimension on the national level and in a North–South perspective in Jutland with regard to the many similarities of the aims and tools. The remaining part of this section will focus on a specific implementation of the knowledge and cluster relevant targets in the regional development strategies summarized in Table 12.1 with point of departure in the regions of Southern Denmark, see map in Appendix 12.2.

In fall 2007, the region of Southern Denmark launched a new cluster initiative aimed not only to support existing clusters (e.g. facilitating spin-off processes from existing firms), but also to spot and support potential clusters and identify germinating or budding clusters.

At the regional level, a strategy for regional development based on the growth model was introduced as a policy guideline by the national business development agency based on four growth drivers. These drivers include (1) innovation, (2) entrepreneurship, (3) information and communication technology and (4) human resources. In an earlier version, welfare service was included as well (Copenhagen Economics, 2004; Cornett, 2009, pp. 399–404). This has been transferred into a working program focused on areas of priority (see Table 12.1). The region of Southern Denmark has launched a program focused on the six headlines found in Table 12.1, including a cluster strategy. The overall aim is to facilitate a regional development focused on economic growth, housing, service and culture. This broader part of the strategy also contributes to maintaining regional attractiveness for business and industry and preventing crucial mobile location factors (i.e. human capital) or knowledge anchors from moving out of the regions with the consequence of eroding the location base for innovative and fast-growing industries and services. This is a major problem, particularly in the more peripheral localities of the region.

The cluster development initiative is a central building block in the development policy of the region of Southern Denmark. The remainder of this section will detail their cluster development initiative. The basic assumption of the initiative is that clustered industries can contribute to economic growth and development. They usually perform better than non-clustered industries (Erhvervs og Boligstyrelsen, 2003). For this reason, cluster policy has become a popular element in business development strategies in the new Danish regions despite academic studies that question the belief (Engelstoft et al., 2006). The starting point of the cluster strategy was sketched by the growth forum secretariat at a public launch in Vejle in January 2007 (Vækstforum Region Syddanmark, 2007). A summary of targets and the overall idea of the concept are outlined in Figure 12.2.

From a growth perspective, the idea is that mature clusters, as important as they may be for a region, usually have limited growth potential. From a business development perspective they can contribute as anchors for new firms or as sources for spin-offs (see Figure 12.1). Here, the main task for the business development system is to facilitate the cluster within the framework of traditional business services offered and to increase the performance of the involved companies (note dotted line in Figure 12.2).

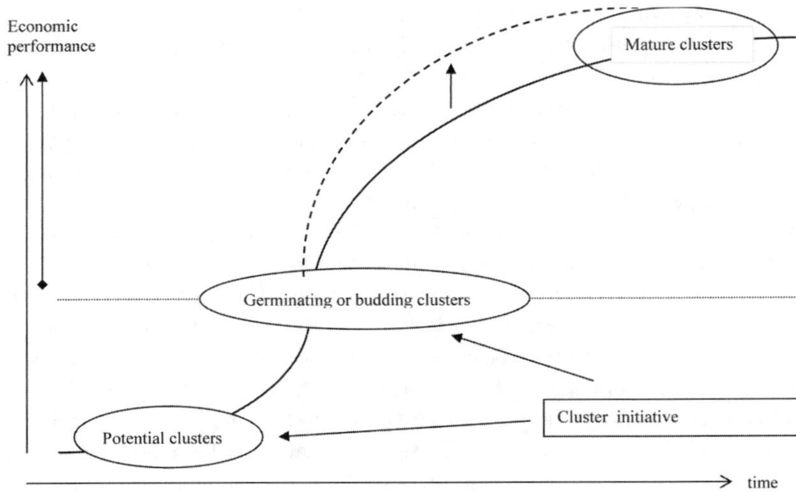

Based on Cornett (2008) and Klyngeudvikling Syddanmark (2009a, p. 18)

Figure 12.2 The cluster initiative

The innovative element of this initiative was to identify potential clusters and to stimulate firms and other involved actors to become germinating or budding clusters. In an explorative study conducted for the region of Southern Denmark by CESFO and IER of the University of Southern Denmark and in cooperation with Force Technology, Green House South and Oxford Research, an attempt was made to map the 'cluster landscape' according to the categories of Figure 12.2. The study identified six mature, 28 budding and 27 potential clusters in the region (Klyngeudvikling Syddanmark, 2009a, p. 24). The result included 88 cluster initiatives. Based on interviews this list was classified into the above-mentioned groups and the 28 budding clusters were investigated in more detail. As a result of this scanning process 10 budding or germinating clusters were chosen for in-depth investigation (see Table 12.2). The method used was rather eclectic, based on a survey among business development agencies, organizations and involved actors. For details on the survey see Klyngeudvikling Syddanmark (2009b).

The next challenge was to develop appropriate tools to support their development. With regard to selection of potential clusters, the task was even more difficult since this can primarily be done in various ways. Studies of successful clustering processes in other regions suggest that choice be determined through investigation of the local industrial base. It is important to stress that the search process should not necessarily be limited to traditional spatial-based clusters

Table 12.2 Identified promising potential clusters

Highest potential	Large potential	Some potential
Steel cluster	Energy cluster	Entertainment cluster
Plastic cluster	Food cluster	Environment cluster
	Transportation cluster	Health cluster
	Tourism cluster	Security cluster

Source: Klyngeudvikling Syddanmark (2009a, pp. 234–39)

but must include virtual or non-spatial patterns of cooperation. One example of a non-spatial cluster is the so-called 'AluCluster' located in the south-western part of the region. Geographically, this cluster includes the Alumina industry developed around the city of Tønder. The 'AluCluster' is an organization providing service for this industry with members found throughout the country. Regardless of the name, it can be argued that this cluster is more a network-facilitating organization than a cluster according to the definition cited earlier:

> AluCluster's mission is to meet the need for highly qualified consultancy now and in the future within practical utilization of aluminium and thereby become the preferred partner within consultancy and development of aluminium solutions. AluCluster consists of two units; a business unit and an innovation unit. The business unit's main objective is to sell consultancy within our competence areas such as product and process development, design, material technology, corrosion, surface treatments, joining including gluing etc. AluCluster's business unit offers companies adapted aluminium courses within all levels in the companies. The innovation unit's main objective is to maintain, develop and establish new workplaces within the aluminium area. (AluCluster, 2008)

Since the cluster policy of the region is still not fully operational it is impossible to present an outline of program implementation. The crucial elements from a conceptual point of view are:

- the identification of sustainable potential clusters in the region or of firms engaged in potential non-spatial clusters;
- advice and organizational input for firms and institutions engaged in cluster development processes;
- provide input for the decisions of the growth forum regarding cluster policy and the priorities in business development policy.

The immediate result of the survey was the decision by the growth forum to support the cluster development initiative in a rather limited extent.

The recommendations of the cluster report can be summarized under the following headings (Klyngeudvikling Syddanmark, 2009a, p. 218): cluster facilitator, clusterhouse, micro-finance facility, dialogue with R&D institutions and marketing initiatives. In particular the third recommendation is important in the context of this discussion. This recommendation underscores that the cluster initiative is not an isolated instrument; rather, it addresses several of the targets of the regional development strategy (see Table 12.1). Regardless, only rather limited resources were dedicated to the program during the first discussions in the Growth Forum Spring 2009 (Vækstforum Region Syddanmark, 2009). These resources included:

- micro-finance facility 1 million DKK per year for three years;
- 2009 and 2010: 0.5 million of the above amount reserved for greenhouses to support food and transport cluster development;
- regional business development department has to provide an overview of cluster actors and relations (e.g. Regional Greenhouse, Reg X);
- a successful knowledge-based cluster development policy depends on financial support and on the ability to identify the most promising initiatives.

From a scientific and research perspective, the development of appropriate methods for scanning potential clusters and the screening of the potential for identified 'clients' is the most critical part of the program together with the creation of appropriate measures to support the realization of the identified potential and their transformation into an active or germinating cluster. This points back to the classification developed by Pickernell et al. (2007), and it seems that in particular the clusters and network frameworks labelled 'Social network' and 'Virtual organization', with their focus on knowledge and learning in a relational perspective, provides a useful starting point for an innovative and knowledge-oriented cluster development policy.

From an institutional context, the main challenge was to avoid the influence of vested interests of local authorities or industries due to the potential economic aid available, and to avoid being trapped in what can be called a 'wishful thinking' cluster. From a knowledge economy perspective, it must be noted that some of the most knowledge-intensive examples are found in the category of budding clusters or are considered to have lower potentials in an economic growth and development perspective for the region.

SUMMARY AND PERSPECTIVES

From a conceptual point of view it seems possible to create a business environment that facilitates the appearance of a business community characterized by features traditionally connected with industrial districts or clusters that are both traditional and knowledge-based. This is accomplished by combining broadening elements added by Pickernell et al., 2007 and Porter (2003). A proactive supply-side policy – based on triple helix inspired concepts – is particularly important in the latter case. The conceptual discussion and the empirical example presented in this chapter have stressed several alterations necessary to use a proactive cluster developing strategy. These include the following:

- The beneficial effects of traditional industrial districts do not necessarily require physical proximity. Virtual systems based on shared knowledge and technology become feasible (see Pickernell et al., 2007).
- The concept of clusters has to be extended to cover the whole range of the business environment (i.e. the 'triple helix' framework, but also the advisory system and public support providers, etc.), and the relevant functional linkages have to be specified, particularly in the virtual case.
- The beneficial effects of proximity and shared skills, traditionally the source of the competitive advantage of industries located in industrial districts, will be replaced or at least supplemented by advantages based on virtual systems and the availability of relevant consultants and advisory agencies. The latter is particularly important for small and medium-sized enterprises.

This leads to the final and most important part of the chapter; namely to answer the question of whether or not the creation or stimulation of clusters is a feasible objective for regional development policy in a knowledge economy context. The crucial issue is how to manage a cluster-developing policy that both solves the regional development purpose (i.e to create positive societal externalities) and at the same time secures the self-interests of the involved firms and agencies. Furthermore, the tendency toward institutionalization of cluster initiative leads to a need to provide a sustainable financial base, also after the phase of creating the framework. The above-mentioned 'AluCluster' example proved the problem of bridging the gap between non-profit cluster network functions and consultancy work to finance the activity.

Based on the conceptual and theoretical aspects discussed, and the reported policy proposal from the region of Southern Denmark, an affirmative answer to the question whether a cluster-stimulating policy can encourage the development toward a more knowledge and innovation-based economy or not, could be possible. The crucial challenge is probably not the support of existing or budding clusters, but the development of tools to identify potential clusters in the regional business environment. Future research in this area needs to include:

- further conceptual work on the nature of 'space-less' or virtual clusters and how they relate to business and industry or to knowledge institutions in a particular region;
- the development of appropriate tools to deal with how this type of 'clustering process' can contribute to beneficial spin-offs or positive externalities in a region;
- elaboration on the concept of 'potential clusters' and how to identify the most promising 'clusters' in a specific context.

With regard to the latter, a comprehensive mapping of the linkages between the relevant entities within and outside the investigated region is needed to identify all relevant potential beneficial spillovers. These extend beyond the interrelationships of existing actors to include new startups and entrepreneurs in the cluster concept. This is of particular importance in the political prioritization of knowledge-based entrepreneurship. In a survey of literature on knowledge spillover theory and spatial clusters, David Audretsch suggests rather mixed results and states: 'virtually none of these studies provided a theory linking knowledge spillovers to new-firm startup activity, nor did any of these studies provide a measure of knowledge spillovers' (Audretsch, 2006, p. 185). Nevertheless, some empirical evidence of the importance of knowledge as a driver of economic growth was found in a study based on German data (ibid):

> evidence shows that startup activity tends to cluster geographically around sources of new knowledge, such as research and development (R&D) investments by firms and research undertaken at universities. Their findings provide compelling support for the *Knowledge Spillover Thesis of Entrepreneurship* in that entrepreneurial activity is systematically greater in locations with a higher investment in knowledge and new ideas. (Audretsch, 2006, p. 185)

A final and still open question involves how to solve the problem that economic activity and prosperity seem to become more centralized, leaving the rural and peripheral regions behind. Neither the theoretical nor empirical discussion presented in this chapter provides conclusive answers as to how the build-in tendency in modern economic development toward geographical concentration in the knowledge economy can be reversed or changed in a way that also facilitates activities in the peripheral locations.

NOTE

1. The concept 'industrial district' is not without its ambiguities. Ann Markusen (1996) distinguishes between four types: the Marshallian NID with Italian variants, the hub and spoke

district, the satellite industrial platform and the state-centered district (p. 296). The issue of conceptualization has been developed further by Pickernell et al. (2007) with special attention on economic development, see p. 216.

REFERENCES

AluCluster (2008), 'Knowledge and technology center for aluminium', retrieved on 13 May 2011 from www.alucluster.com.

Andersen, P.H., A. Bøllingtoft and P.R. Christensen (2006), 'Erhvervsklynger under press: globaliseringens indflydelse på udvalgte danske erhvervsklynger', *Ministry of the Environment*, Copenhagen.

Audretsch, D.B. (2006), 'The knowledge spillover theory of entrepreneurship and spatial cluster', in Iréne Johansson (ed.), *Innovations and Entrepreneurship in Functional Regions,* papers presented at the 8th Uddevalla Symposium and the 8th McGill International Entrepreneurship Conference, 15–17 September 2005, Uddelvalla, Sweden, research reports 2006:01, University West, Trollhättan; Sweden, pp. 177–88.

Bergman, E.M. and E.J. Feser (1999), 'Industrial and regional clusters: concepts and comparative applications', *West Virginia Web Book of Regional Science* (http://www.rri.wvu.edu/WebBook/Bergman-Feser/chapter2.htm).

Cheshire, P. and E.J. Malecki (2004), 'Growth, development, and innovation: a look backward and forward', *Papers in Regional Science*, **83** (1), 249–67.

Copenhagen Economics (2004), *Regionernes konkurrenceevene*, Copenhagen Economics and Inside Consulting Report for the Ministry of the Interior and Health, January 2004.

Cornett, A.P. (2009), 'Aims and strategies in regional innovation and growth policy – a Danish perspective', *Entrepreneurship and Regional Development*, **21**(4), 399–420.

Cornett, A.P. and N.K. Sørensen (2007a), 'International vs. intra-national convergence in Europe: an assessment of causes and evidence', paper prepared for the 47th Congress of the European Regional Science Association, 29 August–2 September 2007, Paris/Cergy-Pontoise, France.

Cornett, A.P. and N.K. Sørensen (2007b), 'Regional restructuring: creating new advantages or reinforcing old disadvantages: a critical assessment of development policy impacts of the regional and municipality reform in Denmark', in Iréne Johansson (ed.), *Institutions for Knowledge Generation and Knowledge Flow: Building Innovative Capabilities for Regions*, papers presented at the 10th Uddevalla Symposium, 14–16 June 2007, Uddevalla Sweden, research reports 2007:04, University West, Trollhättan, Sweden, pp. 347–65.

Danske Regioner and Kommunernes Landsforening (2007), 'Regional vækst og udvikling: et fælles ansvar Samarbejde om vækstfora, væksthuse og regionale udviklingsplaner', Copenhagen.

Engelstoft, S., C. Jensen-Butler, I. Smith and L. Winther (2006), 'Industrial clusters in Denmark: theory and empirical evidence', *Papers in Regional Science*, **85** (1), 73–97.

Erhvervs og Boligstyrelsen (2003), 'Kortlægning af danske kompetenceklynger', Copenhagen.

Etzkowitz, H. and L. Leydesdorff (2000), 'The dynamics of innovation: from national systems and "Mode 2" to a triple helix of university-industry-government relations', *Research Policy*, **29**, 109–123.

Gordon, I.R. and P. McCann (2000), 'Industrial clusters: complexes, agglomeration and/or social networks?', *Urban Studies*, **37** (3), 513–32.

Illeris, S. (1999), 'Serviceerhverv, nærhedsbehov og regional udvikling', paper presented at the biannual meeting of the Danish section of NS-RSA, Svinkløv, September 1999.

Klyngeudvikling Syddanmark (2009a), 'Spirende klynger i Region Syddanmark Status og politiske anbefalinger vedrørende klyngeudvikling', retrieved on 8 September 2009, from http://vaekstforum.regionsyddanmark.dk/wm220760.

Klyngeudvikling Syddanmark (2009b), 'Spirende klynger i Region Syddanmark Anvendt metode ved kortlægning og analyse af klynger', retrieved on 8 September 2009 from http://vaekstforum.regionsyddanmark.dk/wm220760.

Markusen, A. (1996), 'Sticky places in slippery space: a typology of industrial districts', *Economic Geography*, **72** (3), 293–313.

Marshall, A. (1947), *Principles of Economics*, 8th edn, London: Macmillan and Co. Ltd.

Malmberg, A. and P. Maskell (2002), 'The elusive concept of localization economies: towards a knowledge-based theory of spatial clustering', *Environment and Planning A*, **34**, 429–49.

Ministry of Economic and Business (2006), *Regionalpolitisk vækstredegørelse 2006*, 20 May, Copenhagen.

Ministry of Industry (1998), *Erhvervsredegørelse 1998*, Statens Information, Copenhagen.

Nordregio (2009), 'The new administrative structure of Denmark after the structural reform in 2007', retrieved on 24 November 2009 from http://www.nordregio.se/.

OECD (Organisation for Economic Co-operation and Development) (2001), *The New Economy: Beyond the Hype the OECD growth report*, Paris: OECD.

Pickernell D., P.A. Rowe, M.J. Christie and D. Brooksbank (2007), 'Developing a framework for network and cluster identification for use in economic development policy-making', *Entrepreneurship and Regional Development*, **19**, 339–58.

Porter, M.E. (1998), *On Competition*, Boston, MA: Harvard Business Review Book.

Porter, M.E. (2003), 'The economic performance of region', *Regional Studies*, **37** (6 and 7), 549–78.

Signorini, L.F. (1994), 'The price of prato, or measuring the industrial district effect', *Papers in Regional Science*, **73** (4), 369–92.

Steiner, M. (2002), 'Regional policy in transition institutional challenges and strategic perspectives', paper for the 15th European Advanced Studies Institute in Regional Science, Emerging Market Economies and European Economic Integration, 10–16 June 2002, Eksjö.

Vækstforum for Region Hovedstaden (2007), 'Partnerskaber for viden, vækst og velfærd – erhvervsudviklingsstrategi', Region Hovedstaden Januar 2007. Retrieved on 13 May 2011 from http://www.regionh.dk/NR/rdonlyres/6604AC23-6EE9-44B6-9B6D-68C903BF7266/0/Netversionafestrategien.pdf.

Vækstforum Midt (2007), 'Fornyelse og vækst i en international vækstregion Erhvervsudviklingsstrategi 2007–2009 Handlingsplan 2007–2008', Region Midtjylland Viborg. Retrieved on 13 May 2011 from http://www.ebst.dk/file/5765/handlingsplan_midtjylland.pdf.

Vækstforum Region Nordjylland (2007), 'Vækst og Balance: Erhvervsudviklingsstrategi for Nordjylland 2007–10', Aalborg. Retrieved on 13 May 2011 from http://www.rn.dk/NR/rdonlyres/4274541B-A448-4743-8F3A-1F2976FA38A1/0/Erhvervsudviklingsstrategi_komplet_udgave.pdf.

Vækstforum Region Syddanmark (2007), 'Erhvervsudviklingsstrategi 2007–2012', Media Syd, Vejle.
Vækstforum Region Syddanmark (2009) 'Minutes Growth Forum March 31', 2009 Region Syddanmark. Retrieved on 13 May 2011 from http://vaekstforum.regionsyddanmark.dk/wm276709.
Vækstforum Sjælland (2007), 'Strategi 2007–2010: Vækstforum Sjællands erhvervsudviklingsstrategi 2007–2010', Retrieved Spring 2008 from www.regionsjaelland.dk, currently not available.

APPENDIX 12.1: THE GROWTH FORUM AND ITS PARTNERS OF COOPERATION

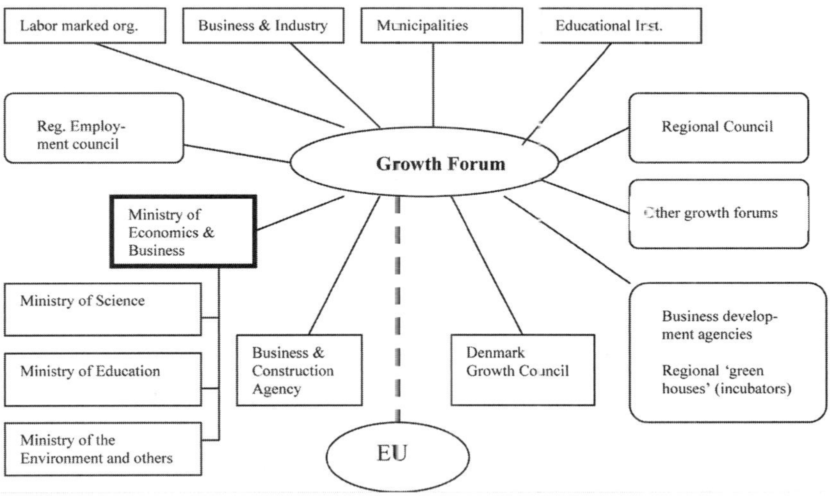

Source: Adapted from Danske Regioner and Kommunernes Landsforening, 2007

APPENDIX 12.2: THE NEW ADMINISTRATIVE STRUCTURE OF DENMARK AFTER THE STRUCTURAL REFORM IN 2007

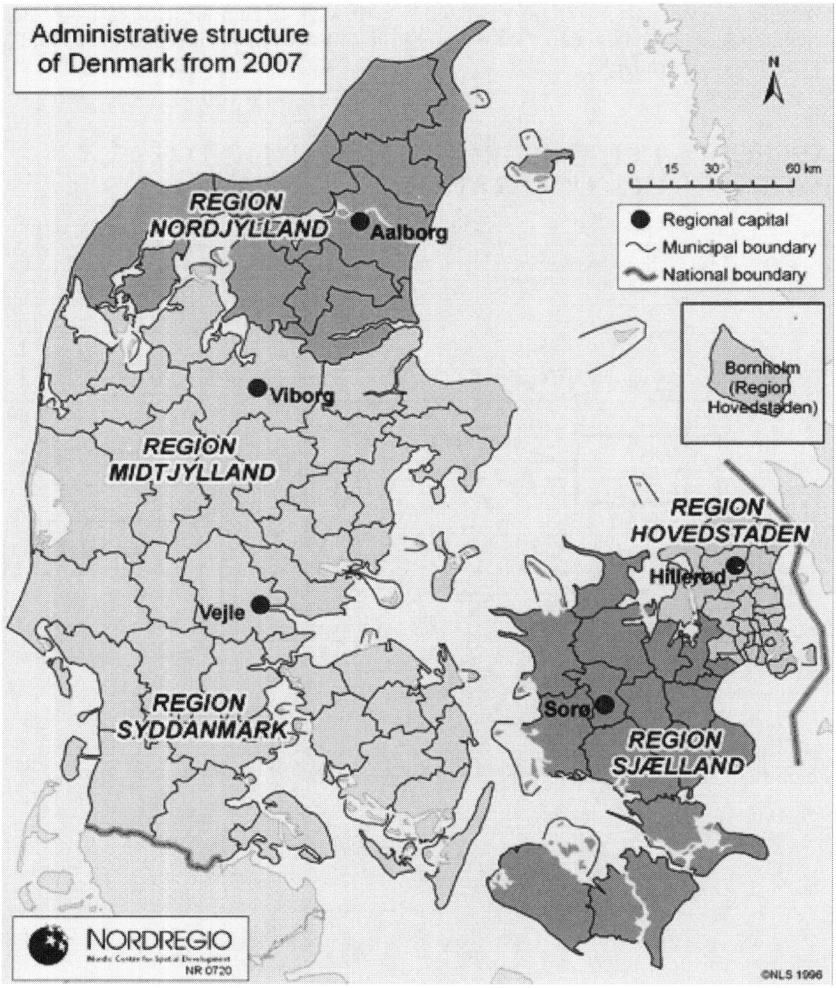

Source: Nordregio (2009)

13. International virtual industry clusters and SMEs: early process policy recommendations

Meir Russ and Jeannette K. Jones

INTRODUCTION

The new knowledge-based economy puts forward distinctive contests to individual entrepreneurs, companies, regions (e.g., Statistics Canada, 2005), and industry clusters (e.g. Russ and Jones, 2011). One of the more intriguing challenges encases the complex nature of tacit knowledge. On one hand, the importance of tacit knowledge as an endogenous foundation of growth by business strategists at the individual/entrepreneur and company levels (e.g. Audretsch, 2003) and as the exogenous source of economic growth is now widely received by economists for the regions and nations (e.g. Statistics Canada, 2005). At the same time, the importance of the stickiness of tacit knowledge is increasing (e.g. Audretsch, 2003). This tacitness and the stickiness of knowledge enforce new characteristics on public policy toward economic development (e.g. Johansson et al., 2001).

Can the new information and communication technologies (ICTs) help in resolving this challenge? Counter-intuitively, until recently this was not the situation. The shortened half-life and sticky nature of knowledge greatly enhanced the weight of face-to-face (F2F) interactions which played a major role in business success and wealth creation (e.g. Dohse, 2003), and the importance of learning in regional networks for knowledge development purposes (e.g. Fedderke and Klitgaard, 1998). In addition, the increased pace of new knowledge creation resulted in a higher concentration of industry clusters in major urban areas (Florida et al., 2007), as well as increasing the income gap between the munificent regions and the poor, rural areas (e.g. Russ and Jones, 2011).

This chapter will propose that in order for a public policy to be effective in the new knowledge-based economy, the policy makers need to be able to understand the nature of the life cycle of the industry clusters, the unique nature of tacit knowledge and the effect it has on the specific industry cluster (Cooke, 2007), and the implications of the next generation ICT, specifically Web 2.0, video-conferencing and mobile telecommunication (e.g., Stephens, 2008). Two mega-trends seem to be the major drivers behind this new economic development milieu: the 'decoupling' of the value-added chain, which allows for a major industry revamp (e.g. Enright, 2000), and the younger generation adoption of the social media (e.g. Xenos and Foot, 2008). The importance of the life cycle of the industry clusters will be discussed along with a focus on the impact on the role of the recently presented proposal for the creation of the International Virtual Industry Cluster (IVIC) (Russ and Paterni, 2007).

The chapter will proceed as follows. First, we will introduce the concept of IVICs. Next, we will present the framework of a process model for policy implementation, including specific implications for IVICs. This will be followed by policy conclusions and proposals for next steps.

INTERNATIONAL VIRTUAL INDUSTRY CLUSTER (IVIC)

Recently, Russ and colleagues (Russ and Jones, 2006, 2008; Russ and Paterni, 2007) proposed an alternative governance form that might allow an alternative to the centrifugal trend toward increased urbanization of the mega-regions and the increase of gaps in regional inequality. The proposal includes the concept of international virtual industry cluster (IVIC) as a critical building block of an international virtual mega-region (IVMR). This proposal was based on the conclusions of a recent study performed by the first author in Northeast Wisconsin (Russ, 2006). That study developed and validated a framework and set of indicators for a knowledge-deprived, economically peripheral region, which included the analysis of regional, sub-regional and industry cluster performance and indicators. The authors also suggested that a more in-depth discussion of all the streams of economic development industry cluster's theoretical background was needed to allow practitioners to select the appropriate aspects applicable for their context and strategies (Russ and Jones, 2006). A final recommendation of that study (see Russ and Jones, 2008) suggested also overcoming the lack of knowledge assets in the knowledge-deprived region by the development of a virtual mega-region. Such an approach could compensate for the lack of a critical mass of knowledge assets by bringing together multiple regions, each one with complementary assets. An example of such an attempt is the collaborative initiative currently under way between a region in Northern Italy (Pisa-Lucca) and a Northeast Wisconsin region. Some early indicators of successes might be the recent developments of business, municipal and academic relationships between the two regions. This design responds to Enright's (2003, p. 123) proposal of futuristic structures that would encompass the international business arena and the geographically based industry cluster, support the emergent properties of networks of companies (Lipparini and Lomi, 1999) in a deprived region, minimize the cost of networked organization (Ring, 1999), extend Bathelt's (2008) concept of regional knowledge clusters utilizing local buzz and global pipelines, incorporate Molina et al.'s (1998) proposal for virtual industry cluster in the international arena and allow for employing an underutilized economic and social potential of peripheral regions (Nuur and Laestadius, 2007)

The proposed IVIC framework (see Fig. 13.1 and Table 13.1) extends the notion of industry clusters into the international sphere by using the new broadband ICT as a means to shrink geographic distances and replace it (as a critical dimension) with the notion of cultural/psychological distance. The authors proposed that the presently available ICT (e.g. Guinet, 2003) will soon allow the channel richness (Daft and Lengel, 1986) required for an intense

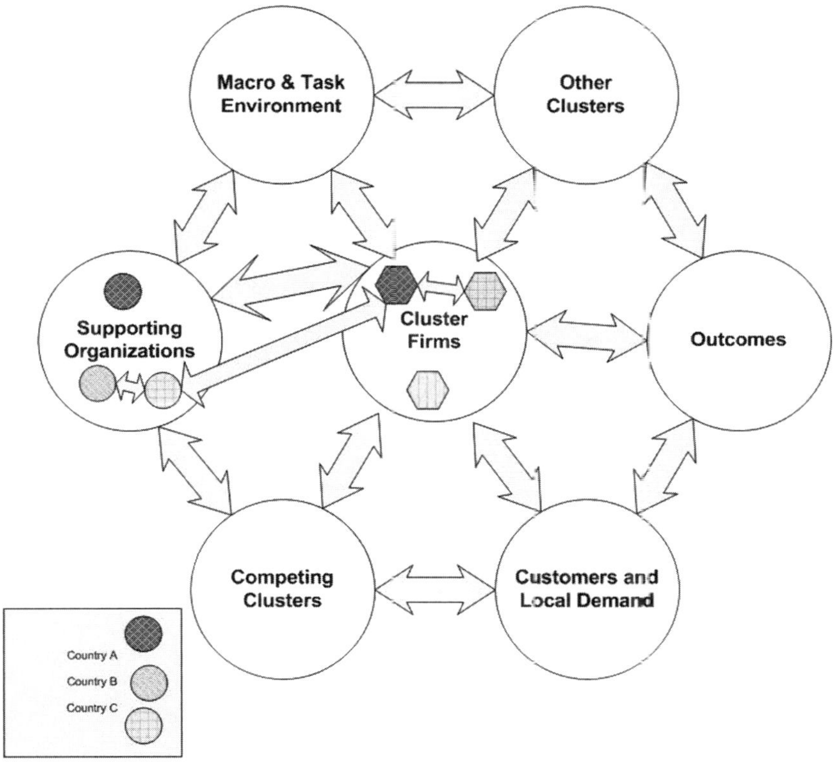

Source: Russ and Paterni (2007)

Figure 13.1 International virtual industry cluster (IVIC) indicators

Table 13.1 Indicators for international virtual industry cluster (IVIC) and public policies

Constructs	Sub-constructs	Indicators
Cluster firms	Linkages between firms[a]	• Partnerships and alliances within IVIC's firm • Involvement in IVIC regional clustering activities • Linkages within the IVIC network
	IVIC life cycle[b]	• Stage of IVIC life cycle • Need for restructuring • New technologies that might revolutionize the IVIC
	Internal awareness[a]	• Internal awareness of cluster members
	External representation[a]	• Existence of shared representation • Involvement in national and international trade associations
External factors	Human capital[b]	• Access to QP[c] • Sources of QP[c] • Distance of QP[c] sources
	Social capital[b]	• Innovativeness relative to competing virtual mega-regions • Quality of local lifestyle • Economic inequality
	R&D capacity[b]	• Contributions of virtual mega-region's institutions to ideas, knowledge and innovation • Availability of technology transfer capacity
	Infrastructure[b]	• Quality of virtual mega-region's transportation infrastructure • Connections to national/international transportation systems Availability of specialized form of infrastructure
	Information infrastructure[b]	• Access to key business information • Quality of communication infrastructure
	Financial and business climate[b]	• Business costs (cost of living, cost of doing business) relative to competing virtual mega-regions
Supporting organizations	Federal and state policies and programs[b]	• Contributions of international, federal and state institutions to ideas, knowledge and innovation • Provisions for venture capital for New Product Development • Targeted inward investment promotions
	Community resources and support[b]	• Adequacy of virtual-mega-region's development support
	Academic resources and support[b]	• Adequacy of academic resources (K-12, vocational schools, colleges, universities) • Virtual mega-regional support

Constructs	Sub-constructs	Indicators
	Suppliers[b]	● Virtual mega-regional availability of materials and equipment
	Services[b]	● Virtual mega-regional availability of business services
	Capital[b]	● Virtual mega-regional availability of capital (specifically venture capital)
Related clusters	Linkages between IVICs[a]	● Number of related IVICs ● Partnerships and alliances between IVICs ● Involvement in inter-clustering activities ● Linkages between IVIC networks
Competitors (competing IVICs)		● Cultural distance of most important cluster's competitor ● Number of most important cluster's competitors
Customers[c]		● Cultural distance of most important cluster customers (and/or potential customers) ● Number of most important clusters of customers (and/or potential customers)
Performance	Size	● Number of firms ● Size of firms
	External reach	● $ sales ● % of export
	Innovation inputs[b]	● R&D spending ● Business development capabilities ● Product development capabilities
	Innovation outputs	● $ revenues from new product development
	Dynamism	● Number of new firms within the IVIC
	Social responsibility	● $ donation
	Recognition	● External recognition by others ● Collective marketing initiatives to attract external actors

Notes:
Based on and modified from Russ and Paterni (2007)
[a] Critical aspects of policy implementation
[b] Critical aspects to consider for planning a policy
[c] QP – skilled employees, experts, inventors

discussion and dialogue across geographical boundaries and necessary to support knowledge spillover among the geographically dispersed actors and as such, allow for the untangling of the geographical and social proximity (e.g. Sooryamoorthy et al., 2008). Those cultural or psychological distances will become of higher importance, at least with regard to the sharing of tacit knowledge and networking of innovations aspects, than physical distances (see example Bathelt et al., 2004). Such cultural distances will have value/norms dimensions as well as temporal dimensions (Ogburn, 1964). Some cultural changes might be required as a prerequisite for enabling the accumulation of intellectual capital in a knowledge-deprived region (Russ and Jones, 2006) (see our discussion, of culture and policy on p. 242).

Russ (2009) suggested the following characteristics of the IVIC.

- IVIC has a mix/blend of geospatial clusters and networks (agglomeration and externalities) that allow it to overcome inherent weaknesses of regional clusters.
- IVIC requires at least two geographical locations (sub-clusters) that have unique value creation propositions. This value proposition must be significantly higher than the traditional alternatives, due to the additional costs and risks involved in the creation and prolongation of the IVIC. The value creation can be revenue opportunities (like customers or experience economics), or cost savings (either low cost of inputs or low cost of transactions), plus knowledge creation (externalities) either within or between the sub-clusters. The high revenues or low cost can result from unique history and/or recombination of tangible and intangible inputs.
- IVIC requires both networking and entrepreneurial capabilities and skills above and beyond the required industry-specific knowledge base.
- IVIC has at least one research university embedded in at least one of the sub-clusters.
- IVIC has direct (physical or virtual) access to knowledge-intensive and financial services.

The international virtual mega-region framework (as mentioned in Russ and Jones, 2008) suggests that infrastructure enablers (technological, government and administrative systems), as well as resources (culture, education, etc.) will have to be in place to ensure successful implementation. This is where policy, due to market failures, will have to come into play (see below).

The new proposed IVIC framework also postulates the importance of the following key players on the international arena (see Russ and Paterni, 2007):

- bridges/gatekeepers/hubs/connectors within and between sub-clusters (for example, professors serving on corporate boards, or business leaders collaborating in educational programs);
- influencers/leaders (industry and/or political leaders that adopt the collaborative perspective/framework that leads local institutions and develops appropriate policies);
- 'boomerangs'/'sea turtles' (key players/individuals returning from a long stay away, bringing back networks and knowledge);

- events (e.g. international conferences, commercial missions/trips);
- agreements (e.g. sister cities, university collaborative arrangements);
- measures (see example in Table 13.1), especially between international players (for example, number of collaborations between universities in one country and companies in another, or the reverse).

For the purposes of this chapter we will focus on the policy implications that are deemed the most critical for the success of the IVIC.

We suspect that IVIC is currently in its infancy: therefore, we have focused our chapter on the embryonic or pre-embryonic stage of clusters' life cycles and on the wishful thinking, policy driven and potential industry cluster (as type of clusters policies) which would be appropriate at the embryonic stage. We also see IVIC as a hybrid of the classic geographically based industry cluster (GBIC) and a network structure.

POLICY RECOMMENDATIONS FOR SUPPORTING SMALL AND MEDIUM-SIZED ENTERPRISES (SMEs) WITHIN IVIC

Introduction

Successful industry clusters increase their network coverage beyond geographical proximity, regardless of ICT, by using social capital when appropriate (Phlippen and van der Knaap, 2007). Saxenian and Hsu (2005) described the development of the connections between the Silicon Valley in the USA and the Hsinchu region in Taiwan as a result of reverse brain migration. In a similar vein, Avnimelech and Teubal (2006) describe the critical role that the returning brain-drain individuals have on the creation of a successful industry cluster (e.g. Biao, 2005) as well as illustrate what, in our opinion, is one of the seeds of IVIC, the network of relationship. The movie industry is a good example of such a network of relationships. Vang and Chaminade (2007) describe and analyse the relationship between and the co-development of the movie industry within the Hollywood cluster and the Toronto cluster. In a similar manner, Hellman (2008) describes the early development of a sub-cluster of the movie industry in Jerusalem, and Parris (2004) describes and analyses the development of the movie cluster in London.

One key to establishing a viable IVIC is that there will be a critical mass of returning individuals such as, for example, the case of Taiwan and the Silicon Wadi of Israel and Silicon Valley. The question here is whether or not government or regional policies can create or accelerate this process. An example of a successful effort is the subsidies that Indian and Chinese governments provided for returning executives, scientists and engineers who intended to start their own business in the industries and regions the government was targeting (e.g. Tan, 2005). This illustrates the impact of the collision of, and the tension between, two major trends, the growing centrifugal impact of globalization (global–local linkages, e.g. Chaminade and Vang, 2006) and the increasing centripetal impact of the growing importance of the localization of (tacit) knowledge. As suggested elsewhere (see Russ and Jones, 2011), the new ICT (e.g. video-presence and Web 2.0) should allow, at least in some cases, for the creation of IVIC. We claim that in the marginal areas (due to market failures) this will happen only when policies intervene. The munificent regions do not need policy intervention; the market forces and players will take care of it one way (e.g. local cluster) or another (e.g. global open source).

The Recommended Process

The recommendation in this chapter outlines the stages (X_i) and some of the policies ($X_i a$) that will be required to support the development of an IVIC. The framework for the process aspect of the policy (based on and modified from Stimson et al., 2005) is presented below (see Table 13.2 and Figure 13.2).

Table 13.2 The stages of the policy process for the IVIC within the learning region perspective

Stage #	Stage	Effected directly by external environment	Effected directly by policy at the different levels
1	Leadership	No	No
2	Culture	Yes	Yes
3	IVIC self-identity	Yes	Yes
4	Vision	No	No
5	Strategy	Yes	Yes
6	Change in capacity and capabilities	No	No
7	Infrastructure	Yes	Yes
8	Mechanism for using and tapping	No	No
9	Markets	Yes	Yes
10	Absorptive capacity	No	Yes
11	Outcomes	Yes	Yes
12	Sustainable development	Yes	Yes

Based on and modified from Stimson et al. (2005)

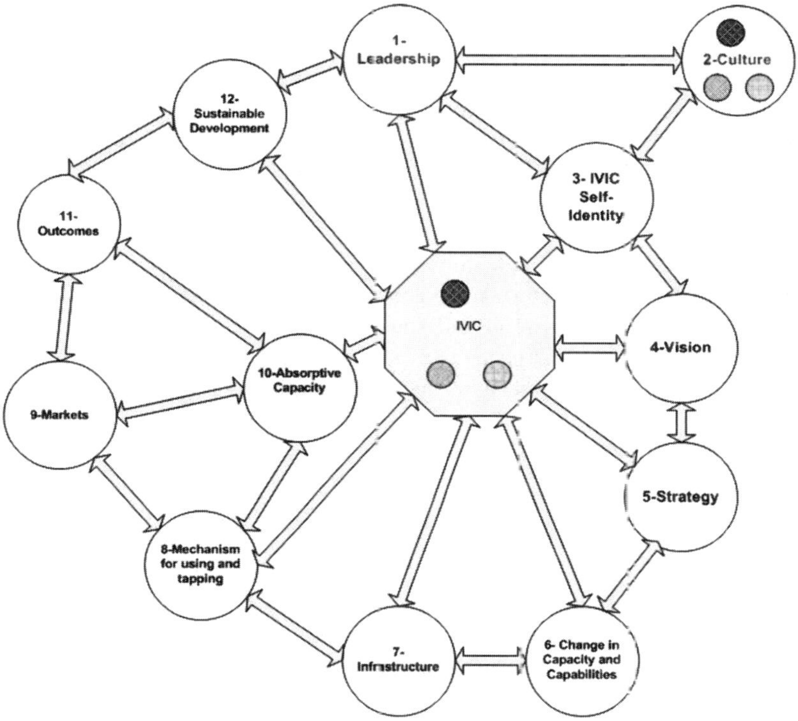

Adapted and modified from Stimson et al. (2005)

Figure 13.2 The policy process framework for the IVIC within the learning region perspective

Leadership

The first, and most important, but not sufficient ingredient for the formation of IVIC is leadership, or what Gerencser et al. (2008) called initiators. Since actors, issues, solutions and chance events, according to the 'garbage can model of decision making', occur continuously and randomly, only when those four streams meet will the appropriate actors' decisions and actions take place (Cohen et al., 1972). Leaders are only one type of actor that plays a key role in the formation of IVICs but their presence is crucial at the early stages of the process. Once IVICs are initiated, the leaders will continue to drive the process by modifying the culture and driving the needed change, by establishing the legitimacy of the IVIC as described below, and by adapting the strategy and goals to the changing circumstances. Who might be the early leaders or the cluster founders, what Stoerring and Christensen (2008) describe as clusterpreneurs, of the IVIC? Possible candidates for leadership are the entrepreneurs of the local clusters (see Feldman et al., 2005, for rationale); specifically, those entrepreneurs that have both local and non-local pre-founding experience (see Simons and Roberts, 2008, for example) and have higher entrepreneurial human capital. Other pools of candidates could be the local government officials engaged in international activities, a pioneering multinational corporation acting as an institutional entrepreneur and/or university personnel involved in international collaborations within the local clusters. The more effective leaders will be the transformational leaders due to their abilities to foster collaboration and trust in the virtual and spatial space (e.g. Strang, 2009), aspects that are crucial at the early stages of new cluster development, regardless of whether or not it is a bottom-up or a top-down initiative.

Culture and Policy

The most difficult to achieve, yet necessary, prerequisite condition for the success of IVICs in a peripheral region is cultural change (as suggested earlier). As Lazerson and Lorenzoni (2005) found, one characteristic of a peripheral region is its higher tendency to have a homogenous culture that resists change. Such culture is, in many cases, one of the causes of the decline and even the failure of industry and/or the region (Copus et al., 2008). Unfortunately, cultural change cannot be mandated. It takes leadership (or a major crisis), at the national level and local level to drive that change. The culture that is needed for successful clusters is one of embracing change and risk, entrepreneurial spirit, collaboration, and embracing growth based on the new knowledge-based economy (e.g. Russ and Jones, 2008).

We accept that cultural change can result from the introduction of new technologies, or from new, top-down-driven legislation, or from failure (e.g. Russ and Jones, 2006). However, there is also the need for some early indication of willingness and readiness for accepting the need for change, at least in the case of a top-down-driven cluster initiative. As such, we will call it the 'readiness for cultural change prerequisite test'. Policy makers who intend to drive the creation of IVIC will have to identify readiness and the presence of the transformational leadership needed for successful initiation of the IVIC (see a rare example discussing this at Stoerring and Dalum (2006, p. 16)).

Public policy can enhance and/or accelerate the pace of cultural change and enhance the probability for success of the top-down policy-driven or bottom-up cluster-driven initiative. Such a supportive policy can also ensure that the appropriate stakeholders participate from the inception and/or increase the probability that trusting relationships are developing and that the probability of organizational failure is minimized. Public policy might influence, indirectly, the culture by providing economic subsidies resulting in individual and organizational relocation and grafting new culture to the region (see illustration in Su and Hung, 2008). One critical aspect of such policy must be support for networking and collaboration between the different constituencies of the cluster.

IVIC Self-identity: Creation and Change and Policy

The leaders need to create the new and socially constructed identity of the IVIC. This could be a 'quantum leap', at least for the early creators of IVIC within the economic development sphere. Two existing organizational structures that can provide a prototype will assist with this step; the 'international virtual teams' and the 'industry clusters'. International virtual teams are frequently used by multinational companies and the technological infrastructure required to support such an organizational form in a peripheral region is becoming available at a reasonable price. Industry clusters are well accepted organizational vehicles within economic development. The new form of IVIC, as such, will have a relatively easy path for acceptance. Such new form creation must take into consideration the legitimization process while being framed in terms of the emerging domain and matching between the nature of the emerging new entity and the participant's identities. Developing the identity of the IVIC should not replace any locally based identity but should be complementary to it and thus minimize the potential for conflict. Two dimensions of the IVIC identity should be considered: the strengths of the identity, and its focus. Two characteristics of the IVIC should be also considered: the dominance of the cluster and the interrelatedness within and between the cluster's constituencies. Since the IVIC is based mostly in the peripheral region, the resources (financial and human) can support probably only a narrow focus and scope of the cluster. The self-identity can be reinforced by using social networks and developing trust.

The next challenge for the leaders will be marshalling the resources needed for and supporting the creation of IVIC in the peripheral region. This is where policy can support the creation and development of IVIC. A formal recognition and support by policy makers will make it easier to attract resources by legitimizing the new organizational form. On the other hand, the ability to attract external and internal resources will make it easier to establish the identity of the IVIC. Such identity-building processes can be lengthy, especially in peripheral regions.

Public policy can also enhance and accelerate the process of establishing the IVIC by legitimizing this organizational form within its geographic and industry context. This can be done directly through a policy statement and by providing budgets, subsidies, legislation, etc. (see example in Su and Hung,

2008) and indirectly by supporting national innovation policies and research and development (R&D) that make it easier on local actors establishing an IVIC. Policy makers will need to recognize the emerging leadership of the IVIC, as well as support, as much as possible, collaboration at the early stages of cluster formation.

Vision

The next step will be creating a vision of the future for the newly created IVIC identity. Such vision scope could be 'global thinking–local acting' (Passiante, 2003, p. 41), meaning a global entrepreneurial opportunity search implemented through local dynamic capabilities. The language and metaphors used for the vision creation will be of critical importance for internal and external effective communication (see example in Rich and Lukens, 2009) and legitimacy purposes, since they will have to operate effectively in culturally diverse environments and in different institutional environments, by attracting financial and human capital resources and by minimizing knowledge asymmetries and focusing on market-oriented innovation. The vision of the IVIC will be shaped by the leading entrepreneurs and their visualization of the opportunity space as well as their understanding of government policies impacting their industry and the economic development in their geographical and psychological proximity.

The model presented so far suggests that the process of the creation and development of IVIC is linear, but in reality, the previously mentioned stages are messy, fuzzy and entail going backward and forward until the new vision is crystallized. A helpful metaphor is a chemical catalyst; an ingredient that can accelerate the process. Such a process accelerator might be an internal charismatic/transformational leader or an expert outsider. Also, as part of the early stages of development, the existing clusters, industries, and so on, should be expected to resist the creation of the IVIC due to organizational inertia and/or because they may potentially lose resources.

Strategy and Policy

The strategy of the IVIC will differ by the stage of the life cycle of the cluster as well as the type. The strategy will define goals and means of achieving them for the IVIC as well as for its different constituencies, government, education, etc. The strategies might include the formal formation of the cluster, including governing bodies, conducting strategic foresight exercises, developing promotional activities, improving on managerial and entrepreneurial skills, promoting investment in infrastructures (e.g. R&D, ICT, etc.) among other options (see Wolfe (2008) for discussion). The strategy will also include the discussion of how to retain existing resources and how to attract needed resources, which will be framed by the identity of the IVIC. For example, the strategy may call for horizontal clustering, vertical clustering, or for a diagonal clustering.

One important question to be answered is: What happens if the cluster is formed as a bottom-up cluster, meaning driven by the industry, or if the cluster is formed as a top-down cluster, meaning driven by the policy makers? If the cluster is formed as a bottom-up, the policy would have a lighter touch. If the cluster is formed as a top-down, the policy would be more direct and specific. Regardless of how the cluster is created, examples of strategies used by the cluster will include using national, state goals, and the means of achieving them, for example subsidies supporting the recognized unique knowledge base that could also be a target for financial public support or private financial investment.

Another area to consider is the appropriate strategy of the cluster in response to the environment and then how policy can support such a strategy. For example, Trippl and Todtling (2008), suggest three alternative strategies for clusters in case of a need for renewal. They recommend for each, the appropriate policy to support the driving entity and the support needed for the knowledge infrastructure and the appropriate networks.

Change in Capacity and Capabilities and Policy

Having the appropriate vision and strategy, the leaders can now engage outsiders that will bring with them knowledge and networks lacking in the peripheral regions, thereby connecting the needed players and resources to overcome the cluster's low organizational IQ (see discussion in Knott, 2008). The outsiders can also increase the capacity to absorb public funds and/or private investments, like venture capital. For example, the outsiders can affect the knowledge of financial investment in knowledge-intensive innovations and the attitude of dealing with the high risk uncertainty of such investments (see example in Copus et al., 2008).

Another example of change in capacity might be the need to restructure the IVIC around knowledge hubs while utilizing diverse local spatial knowledge bases and recombining them, or the need to create new knowledge capabilities in old industrial regions (see example in Trippl and Todtling, 2008). Such a restructuring might create a need for an incremental change, a major change, or a radical change in the capacities, similar to what we would expect in peripheral regions. These changes could also have implications on the needed leadership style.

The needed policy should provide the support to create and/or sustain the capacity and capabilities needed. One example is the instance illustrated by

Su and Hung (2008), of the heavy investment required when creating needed human capital so that the human capital would successfully create a knowledge-intensive cluster. The acquisition of human capital might require up to ten years of continued investment in order to form a successful knowledge-intensive cluster. Academic literature (e.g. Leonard and Swap, 2001) suggests that a ten-year time frame is needed to create the appropriate knowledge base at the individual and organizational level. It can take 15 years in the case of bottom-up clusters and knowledge-intensive clusters (see example in Feldman and Romanelli, 2006). Such a lengthy process might not be easily sustained by the political players, and requires tenacity and perseverance that is rare and difficult to maintain by the political actors, especially in the case of IVIC due to the potential for economic value diffusion in the international arena. Another example is provided by Trippl and Todtling (2008), who suggested having policies supporting adoption, reconfiguration or transformation of knowledge infrastructure.

Infrastructure and Policy

The traditional economic development literature uses the term infrastructure when defining the needed support systems of the IVIC – for example transportation, utilities, and telecommunication infrastructure – but goes beyond that. For the purpose of regional economic development in a peripheral region, we suggest considering the governmental administrative infrastructure as a critical part of the required scope of infrastructure needed (see discussions in Enright, 2000; Russ and Jones, 2008).

The timeframe needed to build different aspects of the infrastructure could be very different, and the changes in the infrastructure will be different as communities might redefine themselves. Also, due to technological changes, the timescales to build the required infrastructure might change.

The specific policies supporting the technological aspect of the cluster's infrastructure are beyond the scope of this chapter, although it is very clear that all transportation, utilities and telecommunication aspects of infrastructure have to be developed. When considering policies regarding governmental administrative infrastructure, the following issues will need specific attention (due to the cross-cultural context of the IVIC):

- One of the most difficult aspects of the policies will be how to create and/or support the trusting relationships between the different constituencies and agents while embedding such trust in a formal structure.
- One of the important administrative infrastructure mechanisms to be put in place should be a conflict-repairing mechanism or process.

- Another administrative mechanism that should be created should support the acceleration of knowledge spillover among the different constituencies of the cluster.
- An additional aspect of infrastructure is the need to provide for random chance events and meetings (what Saxenian, 1994 called 'watering holes'), which can be implemented through support of conferences for a face-to-face or telecommunication infrastructure that will allow for open source collaboration.
- We are only beginning to learn and understand the complex relationship between trust and formal governance (see example in Puranam and Vanneste, 2009).

Mechanism for Using and Tapping

The studies of small world networks suggest that there might be different mechanisms and dynamics operating in networks in close geographical and cultural proximities connecting those networks through bridging ties, and different types of networks. It also seems that different aspects (activities) of value creation networks might benefit from different parts of the small world networks. For example, what might work for innovation and R&D might not necessarily work for customer service or marketing.

In any case, since the structure of the IVIC is more complex than that of the geospatial clusters, it is of crucial importance to minimize the time delays in the learning cycle to make the learning processes effective. An example of learning can be seen below.

We can identify two types of using and tapping mechanisms:

- External mechanisms resulting in improvements of connections to markets, export–import, absorption of knowledge from suppliers and/or customers, and so on. Those mechanisms can be symbiotic or commensalistic in nature. They can also take place in the local milieu or farther away, by using different transfer mechanisms, partially depending on the knowledge type (tacit versus codified) and partially on organizational capabilities, or based on technological (cognitive) proximity between the parties.
- Internal mechanisms resulting in improvements of absorptive capacity by increasing organizational human capital For example, hiring people with expertise and social capital, improving knowledge sharing and cost savings among actors or benefiting from dispersion economics (see discussion in Polenske, 2008) among, within and between the cluster's vertical and horizontal constituencies.

Markets and Policy

The global markets of the IVIC products and services are the final jury for the appropriateness, quality and pricing among many other attributes. The market also provides information and feedback for the clusters' companies.

The markets can be internal to the cluster markets or external to the cluster markets. The markets can also be the result or the cause for companies' collocation while creating the cluster. In either case, the exchanges in the markets (financial or other) will result in financial externalities and also in exchange of data, information and knowledge resulting from financial transactions and social interactions.

The appropriate policy can ease the access to markets, improve the profitability of access to markets, lower the cost of access to markets and lower the risks of access to markets, all within the broader scope of regional and particularly national and international policy guidelines and constraints.

Policy should make it easier for data and knowledge from the external and internal markets to penetrate the cluster and provide for real life feedback to minimize the damage of knowledge filters. On the other hand, subsidies and legislative protection should be minimized and, if used at all, should have limits on the timeframe permitted, so the probability of 'lock-in' (see interesting discussion in Bergman, 2008) will be minimized.

Absorptive Capacity and Policy

The absorptive capacity of the IVIC is the combination of its own cluster absorptive capacity, and its complementary regional R&D capabilities and knowledge bases plus the networking capabilities (social capital) to external knowledge sources. Building the network of knowledge hubs will allow acceleration of new knowledge creations, resulting in a virtual innovation system as a source of sustainable competitive advantage. Cluster companies must also acquire both imitating and emulating learning capacities needed for different stages of the cluster's life cycle, that should allow, by using asymmetric knowledge endowments, creation of spatial monopoly due to scale and scope economics. Such knowledge bases must, at least to a degree, be heterogeneous to allow future changes and adaptation of the cluster to stimulate change when and if needed. The different members of the cluster might develop specializations where some may focus on creating new knowledge. Others may focus on developing the skills of utilizing others' knowledge. Being located in a peripheral region, the ability to absorb others' knowledge is more important, at least at the early stages of the cluster development. Since this capability seems to diminish as the companies mature, companies and the cluster might aspire to continue to invest in such capabilities. It seems that, at least initially, the most important aspects of the IVIC absorptive capacity will be the inter-industry and the intra-industry, while the scientific knowledge might be of less importance. As such, the most appropriate determinants initially will be hiring university graduates and incorporating new and progressive human resource and knowledge management practices (Schmidt, 2005).

Policy typically intends to encourage hiring and investing in R&D employees as investment in human capital. Recently, policy also is encouraging collaboration and partnership, both within and between clusters, especially through informal means.

Outcomes and Policy

The actual outcomes/performances of the IVIC can be compared to the intended outcomes and can also be compared to other regional and/or national sets of indicators that different constituencies might find appropriate.

Some examples of spatial indicators relevant to the knowledge economy that might be used by the IVIC are: ratio of intangible goods and services to tangible goods and services, patenting, level of R&D expenditures, and the use by business of research-based, scientific tools. Raspe and van Oort (2008) introduced a three-dimensional model of regional indicators for the knowledge-based economy that include indicators of knowledge workers, innovation and R&D.

In our earlier papers, we suggested a comprehensive set of outcomes for the IVIC that was driven by specific strategies and could also be used as a benchmark (Russ and Jones, 2006; 2008). Based on our earlier research, a detailed proposal for IVIC's outcome indicators can be seen in Table 13.1. The proposal differentiates between indicators that might be more appropriate for the planning stages of the cluster and those that might be more appropriate for the implementation stages of the cluster. Regardless, choices must be made, since not all the outcomes can be accomplished at the same time.

Policy makers should use the outcomes selected by the IVIC as an input into their future policies. In order to use the outcomes effectively by the IVIC, the appropriate set of outcomes must be reflective and consistent with both of the cluster strategies and the goals of the regional policies.

Sustainable Development and Policy

The self-identity of the IVIC, the resources attracted, and the outcomes will have a significant impact on the continuous and sustainable development of the cluster. The cluster's long term sustainability is not guaranteed, and the most successful clusters can decline or disappear. Clusters can lock in (as a success trap earlier and/or failure trap later) to a technology or customers, and this lock-in might create high exit barriers that will be impossible for the companies and/or the whole cluster to renounce. Another manifestation might be a complete metamorphosis of the cluster. The auto industry cluster in Detroit and the Swiss watch cluster would be well-known examples. The threat to sustainable development of the IVIC can be internal or external and in both cases, companies' responses and supporting policies might and should be able to mitigate the situation.

The policy should support the IVIC's independence and continuous adaptation and adjustment by providing the resources needed. The policy should also support avoidance of too much specialization and should support diversification to minimize risk.

Policy's Conclusions

There are aspects of public policy that have the potential to impact the prerequisite for establishing a cluster, and there are those that can sustain, support and accelerate the cluster building after its inception. In Table 13.1 we recommend which policies and goals would be important to consider when planning for the cluster and those that should be considered when implementing and sustaining the growth of the cluster. At the later stages of IVIC life cycle, the renewal of the cluster will be the most critical stage of policy making. The most tortuous stages for policy intervention are the early stages of cluster development due to the fuzzy and messy nature of those stages. Despite that difficulty, or maybe because of it, we see those early stages as the most important for the future success of any IVIC. Unfortunately, in reality, we suspect that presently, the most effective policies were insinuated only at the later stages of cluster development.

There is another important aspect of public policy and that is that different aspects of policy may not align themselves 'cleanly' into the process model presented in this chapter, and that the different stages (as previously mentioned) might be intertwined, chaotic, and indistinct which will make their implementation more complex.

What industries might be of relevance for policy makers as the 'best' candidates for IVIC? We would recommend looking into optics/photonics, instruments manufacturing and motion picture/film making/movies due to the small size of the companies, high industry fragmentation, short product life cycle, high human capital intensity and high relative value of the knowledge embedded in the products. Other candidates might be science and engineering clusters that provide specific services to a cluster of customers within a specific geographical scope. Lastly, industries that could benefit from dispersion economics (e.g. metalworking sector) might be valid candidates. We strongly recommend that policy makers and economic development consultants that work and have impact on the above-mentioned clusters, consider the policy implications on the different constituencies of the IVIC (as listed in Table 13.1) and the stages (as summarized in Table 13.2) and utilize the indicators we suggested in Table 13.1 to manage and measure the effectiveness and efficiencies of their proposed policies and recommendations.

For policy makers who want to effect specific value-added functions, the most fruitful stages will be at the later stage of IVIC development and in the more mature industries.

NEXT STEPS

Much more research will be required when and if we are able to identify a real-life example of IVIC within an IVMR. Some of the questions to consider at the early stages of the research would be:

- What industries might be the 'best' candidates for IVIC? Are the early candidates identified above relevant and appropriate for the new organizational structure proposed, or are there other, better suited candidates?
- What functions within the value-added chain will the different players have within the IVIC? For example, at the early stages of product innovation, close collaboration between friends that are located across structural holes (geographically and functionally) seems to create high added value (e.g. Bell and Zaheer, 2007). Would this also be the case here?
- What are the specific aspects of psychological distance that play a key role in the formation (or prevention) of IVIC? For example, what role does trust (or lack of trust) play within the different cultures? What interaction is there between the different types of relationships within the players (see example in Bell and Zaheer, 2007) and what impact might the psychological distance have on knowledge sharing?
- Can policy indeed support the creation and maintenance of IVIC? If yes, which aspects of policy will be more effective at which stage of the life cycle of IVIC? Can government support of venture capital firms at the embryonic stage of IVIC creation be effective, especially venture capital firms that are located centrally within their industry networks?

ACKNOWLEDGEMENTS

Earlier versions of this manuscript were presented at the 5th International Conference on Small and Medium Sized Enterprises: Management – Marketing – Economic Aspects; 11–14 August 2008, Athens, Greece and at the 40th MCRSA Annual Conference, Milwaukee, 28–30 May 2009. The first author wishes to acknowledge the Frederick E. Baer Professorship in Business for partial financial support. The authors wish to thank Kelly Anklam for her assistance in editing the chapter and to Dafna Schwartz for her comments. Finally, a special thank you goes to the editor of this book for his thorough feedback and insightful comments. As always, the authors take the responsibility for any mistakes.

REFERENCES

Audretsch, D.B. (2003), 'Globalization, innovation and the strategic management of places', in J. Brocker, D. Dohse and R. Soltwedel (eds), *Innovation Clusters and Interregional Competition*, Berlin: Springer-Verlag, pp. 11–27.

Avnimelech, G. and M. Teubal (2006), 'Creating venture capital industries that co-evolve with high tech: insight from an extended life cycle perspective of the Israeli experience', *Research Policy*, **35**, 1477–98.

Bathelt, H. (2008), 'Knowledge-based clusters: regional multiplier models and the role of "buzz" and "pipelines"', in C. Karlsson (ed.), *Handbook of Research on Cluster Theory*, Cheltenham, UK and Northampton, MA, USA: Edward Elgar Publishing, pp. 78–92.

Bathelt, H., A. Malmberg and P. Maskell (2004), 'Clusters and knowledge: local buzz, global pipelines and the process of knowledge creation', *Progress in Human Geography*, **28**, 31–56.

Bell, G.G. and A. Zaheer (2007), 'Geography, networks, and knowledge flow', *Organizational Science*, **18** (6), 955–72.

Bergman, E.M. (2008), 'Cluster life-cycles: an emerging synthesis', in C. Karlsson (ed.), *Handbook of Research on Cluster Theory*, Cheltenham, UK and Northampton, MA, USA: Edward Elgar Publishing, pp. 114–32.

Biao, X. (2005), 'The international mobility of the highly skilled and development: a comparison of China and India', paper presented at the International Symposium on International Migration and Development, Turin, Italy, 28–30 June 2006. Retrieved on 27 June 2008, from http://www.un.org/esa/population/migration/turin/Turin_Statements/XIANG.pdf.

Chaminade, C. and J. Vang (2006), 'Innovation policy for small and medium size SME in Asia: An innovation systems perspective', in H. Yeung (ed.), *Handbook of Research on Asian Business*, Cheltenham, UK and Northampton, MA, USA: Edward Elgar Publishing.

Cohen, M.D. J.G. March and J.P. Olsen (1972), 'A garbage can model of organizational choice', *Administrative Science Quarterly*, **17**, 1–25.

Cooke, P. (2007), 'Regional innovation systems, asymmetric knowledge and the legacies of learning', in R. Rutten and F. Boekema (eds), *The Learning Region: Foundations, State of the Art, Future*, Cheltenham, UK and Northampton, MA, USA: Edward Elgar Publishing, pp. 184–205.

Copus, A., D. Skuras and K. Tsegenidi (2008), 'Innovation and peripherality: an empirical comparative study of SMEs in six European Union member countries', *Economic Geography*, **84** (1), 51–82.

Daft, R.L. and R.H. Lengel (1986), 'Organizational information requirements, media richness and structural design', *Management Science*, **32** (5), 554–71.

Dohse, D. (2003), 'Taking regions seriously: recent innovations in German technology policy', in J. Brocker, D. Dohse and R. Soltwedel (eds), *Innovation Clusters and Interregional Competition*, Berlin: Springer-Verlag, pp. 372–94.

Enright, M.J. (2000), 'The globalization of competition and the localization of competitive advantage: policies toward regional clustering', in N. Hood and S. Young (eds), *The Globalization of Multinational Enterprise Activity and Economic Development*, Basingstoke: Macmillian Press, pp. 303–31.

Enright, M.J. (2003), 'Regional clusters: what we know and what we should know', in J. Brocker, D. Dohse and R. Soltwedel (eds), *Innovation Clusters and Interregional Competition*, Berlin: Springer-Verlag, pp. 99–129.

Fedderke, J. and R. Klitgaard (1998), 'Economic growth and social indicators: an exploratory analysis', *Economic Development and Cultural Change*, **46** (3), 455–89.

Feldman, M. and E. Romanelli (2006), 'Organizational legacy and the internal dynamics of clusters: the US human bio-therapeutic industry, 1976–2002', paper presented at the 2006 DRUID Winter Conference. Retrieved 8 March 2009, from http://www.druid.dk/uploads/tx_picturedb/dw2006-1725.pdf.

Feldman, M.P., J. Francis and J. Bercovitz (2005), 'Creating a cluster while building a firm: entrepreneurs and the formation of industrial clusters', *Regional Studies*, **39**, 129–41.

Florida, R., T. Gulden and C. Mellander (2007), 'The rise of the mega-region', The Martin Prosperity Institute, University of Toronto, October. Retrieved 28 June 2008, from http://www.rotman.utoronto.ca/userfiles/prosperity/File/Rise.of.%20the.Mega-Regions.w.cover.pdf.

Gerencser, M., R. Van Lee, F. Napolitano and C. Kelly (2008), *Megacommunities: How Leaders of Government, Business and Non-Profits Can Tackle Today's Global Challenges Together*, New York, NY: Palgrave Macmillan.

Guinet, J. (2003), 'Drivers of economic growth: the role of innovative clusters', in J. Brocker, D. Dohse and R. Soltwedel (eds), *Innovation Clusters and Interregional Competition*, Berlin: Springer-Verlag, pp. 150–60.

Hellman, Z. (2008), 'Animating Jerusalem's center', *The Jerusalem Report*, 10 June, 32–34.

Johansson, B., C. Karlsson and R.R. Stough (eds) (2001), *Theories of Endogenous Regional Growth: Lessons for Regional Policies*, Berlin: Springer.

Knott, A.M. (2008), 'R and D/Returns causality: absorptive capacity or organizational IQ', *Management Science*, **54** (12), 2054–67.

Lazerson, M.H. and G. Lorenzoni (2005), 'The firms that feed industrial districts: a return to the Italian source', in S. Breschi and F. Malerba (eds), *Clusters, Networks and Innovation*, Oxford: Oxford University Press, pp. 169–98.

Leonard, D.A. and W. Swap (2001), 'The value of "been there, done that"', in F. Hesselbein, M. Goldsmith and I. Somerville (eds), *Leading for Innovation: And Organizing for Results*, San Francisco: Jossey-Bass Publishers.

Lipparini, A. and A. Lomi (1999), 'Interorganizational relations in the Modena biomedical industry: a case study in local economic development', in A. Grandori (ed.), *Interfirm Networks: Organizational and Industrial Competitiveness*, London: Routledge, pp. 120–50.

Molina, A., S. Ponguta, C.F. Bremen and W. Eversheim (1998), 'Framework for global virtual business', *Agility and Global Competition*, **2** (3), 56–69.

Nuur, C. and S. Laestadius (2007), 'Stuck in the middle? A case study of the underutilized potential in peripheral regions in the developed countries in the age of globalization', *Journal of Rural and Community Development*, **2** (2), 44–63.

Ogburn, W.F. (ed.) (1964), *On Culture and Social Change*, Chicago, IL: The University of Chicago Press.

Parris, S.J. (2004), 'London's film and television industry "cluster": a locally interdependent phenomenon, or a space of intersection of wider networks?', dissertation, University of Leicester. Retrieved 6 June 2008, from http://www.simonparris.postgrad.shef.ac.uk/London's%20film%20and%20television%20industry%20'cluster'%20S%20J%20Parris%202004.pdf.

Passiante, G. (2003), 'Industrial clusters in the net-economy: empirical evidence and some theoretical approaches', in G. Passiante, V. Elia and T. Massari (eds), *Digital Innovation: Innovation Processes in Virtual Clusters and Digital Regions*, London: Imperial College Press, pp. 1–50.

Phlippen, S. and B. van der Knaap (2007), 'When clusters become networks', Tinbergen Institute Discussion Paper, TI 2007-100/3. Retrieved 22 March 2009, from http://www.tinbergen.nl/discussionpapers/07100.pdf.

Polenske, K.R. (2008), 'Clustering in space versus dispersing over space', in C. Karlsson (ed.), *Handbook of Research on Cluster Theory*, Cheltenham, UK and Northampton, MA, USA: Edward Elgar Publishing, pp. 133–49.

Puranam, P. and B.S. Vanneste (2009), 'Trust and governance: untangling a tangled web', *Academy of Management Review*, **34** (1), 11–31.

Raspe, O. and F. van Oort (2008), 'Firm growth and localized knowledge externalities', *Journal of Regional Analysis and Policy*, **38** (2), 100–116.

Rich, T. and L. Lukens (2009), 'Globalizing a function within a company', in J. Salmons and L. Wilson (eds), *Handbook of Research on Electronic Collaboration and Organizational Synergy*, Hershey, PA: IGI Global, pp. 546–59.

Ring, P.S. (1999), 'The costs of networked organization', in A. Grandori (ed.), *Interfirm Networks: Organizational and Industrial Competitiveness*, London: Routledge, pp. 237–62.

Russ, M. (2006), *Regional Economic Development Indicators: A Partnership for Northeastern Wisconsin*, Final Report, submitted to the Executive Board of NEW North, 25 July 2006, unpublished manuscript, University of Wisconsin-Green Bay, WI.

Russ, M. (2009), 'International Virtual Industry Clusters and the International Virtual Mega-Region: A New Strategic Option for Incorporating the SMEs from the Peripheral Regions as a Mode of Entry into the International Business Arena', paper presented at the ISC2009 Conference, Beer Sheva, Israel, 27–29 December 2009.

Russ, M. and J.K. Jones (2006), 'Regional Economic Indicators for a Knowledge Based Economy: A Proposal and Early Findings of a Framework for a Knowledge Deprived Region', paper presented at 53rd annual North American meeting of the Regional Science Association International, Toronto, Ontario, Canada, 16–18 November 2006.

Russ, M. and J.K. Jones (2008), 'Regional economic development indicators for a knowledge-based economy in a knowledge deprived region', *Journal of Regional Analysis and Policy*, **38** (2), 189–205.

Russ, M. and J.K. Jones (forthcoming) (2011), 'Industry clusters, SMEs and public policy: a literature review, and recommendations for peripheral regions', *International Journal of Management and Business*.

Russ, M. and R. Paterni (2007), 'International Virtual Industry Clusters and the International Virtual Mega-Region: A Proposal for a New Paradigm for the

Economic Development in Knowledge Deprived Regions for the New-Knowledge Based Economy', paper presented at the 2007 KMO Conference, Leece, Italy, 10–11 September 2007.

Saxenian, A. (1994), *Regional Advantage: Culture and Competition in Silicon Valley and Route 128*, Cambridge, MA: Harvard University Press.

Saxenian, A. and J. Hsu (2005), 'The Silicon Valley-Hsinchu connection: technical communities and industrial upgrading', in S. Breschi and F. Malerba (eds), *Clusters, Networks and Innovation*, Oxford: Oxford University Press, pp. 235–60.

Schmidt, T. (2005), 'Absorptive capacity – one size fits all? A firm-level analysis of absorptive capacity for different kinds of knowledge', ZEW – Centre for European Economic Research Discussion Paper No. 05-072. Retrieved 22 March 2009, from http://madoc.bib.uni-mannheim.de/madoc/volltexte/2006/1189/pdf/dp0572.pdf.

Simons, T. and P.W. Roberts (2008), 'Local and non-local pre-founding experience and new organizational form penetration: the case of the Israeli wine industry', *Administrative Science Quarterly*, **53,** 235–65.

Sooryamoorthy, R., B.P. Miller and W. Shrum (2008), 'Untangling the technology cluster: mobile telephony, internet use and the location of social ties, *New Media and Society*, **10** (5), 729–49.

Statistics Canada (2005), 'Skills, innovation and growth: key issues for rural and territorial development – a survey of the literature; 1980–2003', Minister of Industry. Retrieved 28 June 2008, from http://www.statcan.ca/english/research/21-601-MIE/21-601-MIE2005076.pdf.

Stephens, R.T. (2008), 'Knowledge management strategies for Web 2.0 integration', in M.D. Lytras, M. Russ, R. Maier and A. Naeve (eds), *Knowledge Management Strategies: A Handbook of Applied Technologies*, Hershey, PA: IGI Publishing, pp. 170–93.

Stimson, R.J., R.R. Stough and M. Salazar (2005), 'Leadership and institutional factors in endogenous regional economic development', *Investigaciones Regionales*, **7**, 23–52.

Stoerring, D. and J.L. Christensen (2008), 'Clusterpreneurs – promotion of high-tech clusters in low-tech regions', Aalborg University, WSP 12-08. Retrieved 5 March 2009, from http://www.business.aau.dk/wp/08-12.pdf.

Stoerring, D. and B. Dalum (2006), 'Mechanism Behind Cluster Emergence – What is the Role of Policy? The Case the Biomedico Initiative in Aalborg, North Denmark', paper presented at the 2006 DRUID Winter Conference.

Strang, K.D. (2009), 'Collaborative synergy and leadership in E-business', in J. Salmons and L. Wilson (eds), *Handbook of Research on Electronic Collaboration and Organizational Synergy*, Hershey, PA: IGI Global, pp. 409–34.

Su, Y.S. and L.C. Hung (2008), 'Spontaneous vs. policy-driven: the origin and evolution of the biotechnology cluster: the case of Bay Area in the United States and Zhangjiang Hi-Tech Park in China'. Retrieved 6 March 2009, from http://www.ceauk.org.uk/2008-conference-papers/Full-paper-Yu-Shan-Su.pdf.

Tan, J. (2005), 'Growth of industry clusters and innovation: lessons from Beijing Zhongguancun Science Park', *Journal of Business Venturing*, **21**, 827–50.

Trippl, M. and F. Todtling (2008), Cluster renewal in old industrial regions: continuity or radical change?', in C. Karlsson (ed.), *Handbook of Research on Cluster Theory*, Cheltenham, UK and Northampton, MA, USA: Edward Elgar Publishing, pp. 203–18.

Vang, J. and C. Chaminade (2007), 'Global-local linkages, spillovers and cultural clusters: theoretical and empirical insights from an exploratory study of Toronto's film cluster', Electronic Working Paper no. 2007/03, CIRCLE, Lund University, Lund, Sweden. Retrieved 4 June 2008, from http://www.circle.lu.se/publications/circle-electronic-working-paper-series.

Wolfe, D.A. (2008), 'Cluster policies and cluster strategies: lessons from the ISRN national study', presented at the annual Policy Day of the Innovation Systems Research Network, Montreal, 30 April. Retrieved 7 March 7 2009, from http://www.utoronto.ca/isrn/publications/NatMeeting/NatSlides/Nat08/Wolfe_Cluster_Policy%20Report_ISRN.pdf.

Xenos, M. and K. Foot (2008), 'Not your father's internet: the generation gap in online politics. Civic life online: learning how digital media can engage youth', pp. 51–70, Cambridge, MA: The MIT Press. Retrieved 22 March 2009, from http://www.mitpressjournals.org/doi/abs/10.1162/dmal.9780262524827.051.

14. The knowledge, learning and innovation process: a poultry cluster in the western region of the state of Parana, Brazil

Jefferson Andronio Ramundo Staduto, Knut Ingar Westeren and Clarissa Pereira Junqueira

INTRODUCTION

The purpose of this chapter is to discuss knowledge, learning and innovation processes of poultry slaughter and processing companies located in the western region of the state of Parana, Brazil. We will also analyse the learning mechanisms of companies, as well as attempt to understand the collection and dissemination flows of innovative activities developed by the companies.

Benefits from clusters are based on the idea that proximity generates advantages to participating companies. External economies resulting from the effects of knowledge flowing into the industry are generated from the concentration of companies that benefit from specialization in a location, also known as MAR transfers, due to work of Marshall (1920), Arrow (1969) and Romer (1986).

The poultry slaughter and processing industry in the western region of Parana state in Brazil takes place in a rural region far from the influence of metropolitan regions and has, therefore, few advantages that are associated with urban economy productive externalities (c.f. Jacobs, 1969). In this particular case, our research suggests that the industry takes advantage of MAR effects.

According to Silva and Silveira Neto (2007), in the MAR approach there are basically three reasons to explain the industrial concentration in a given geographic area: (a) existence of a labor market with specific skills; (b) offers of intermediate inputs at a lower cost and higher variety making the industry more efficient; and (c) information, knowledge, and learning flows more easily due to the geographic concentration of companies and individuals.

External economies that come from agglomerated companies can further be classified in two types characterized as incidental or as attained by planned actions. External incidental economies result from the presence of a group of specialized laborers with specific skills related to agglomerated companies. The presence of specialized raw materials, components and services suppliers, and the dissemination of knowledge, skills and information related to the field of business of local producers, also contribute to these external economies. In addition to incidental external economies, local agents (companies and institutions) may reinforce competitive ability by means of deliberated actions (Suzigan et al., 2003). Such actions often result in the so-called 'collective efficiency'.

By identifying industrial agglomerates, actions designed to guide the cooperation of private agents and request public policy actions to develop dynamic geographically external economies can be proposed. This would result in the creation and consolidation of local competitive advantages seeking to increase the companies' efficiency (Besen and Delgado, 2005).

KNOWLEDGE, LEARNING AND INNOVATION

Knowledge is essential to develop competitive advantages in companies, and it is related to the generation of innovative capability and the ability to use better technologies. Furthermore, it has a close relationship to cognitive ability for the interpretation and processing of information. A more detailed discussion of the knowledge concept is given in Chapter 2 in this book. The formation of knowledge occurs from interactions and interchanges between individuals carrying their own values and beliefs. For formation of knowledge inside organizations, all the specificities in the relationships within companies and among companies, as well as their relationships with other economic, social and institutional agents, must be observed. This question is taken up in several chapters of this book, see for example Chapter 4. These relationships may come from all levels, from local to international (Lemos, 2000).

Learning is looked at as a process in which companies develop their base of knowledge, and it can be understood as a cumulative process that continually develops the company's stock of knowledge. Knowledge and learning have a central role in the development of companies, and consequently society. Thus, technological dynamics can be understood as a continuous process of adopting innovations that depend on the accumulation of knowledge through the learning process. The study of local productive arrangements is justified as a unit of analysis, since it is considered to be an interactive space of learning. In general, the different learning forms of companies can be classified into two major classes: internal learning and external learning. Internal sources of learning are attained from the experience of the company, in design, development, production and marketing of a product. External sources of learning come from suppliers, consumers, universities, research centers, institutes and others (Malerba,

1992). In the empirical part of this chapter we discuss which type of learning process is most relevant for the companies. We use a classification suggested by Rosenberg (1982), where internal learning uses the following three categories:

- learning-by-using is acquired through experience or use and is related to production and to the company's adaptation of new technologies incorporated by new assets – which make viable the introduction of product improvements;
- learning-by-doing is related to the productive process of the company, and generates a continuous flow of modifications and innovations in processes and products. It is also vital to the development of skills in different production stages;
- learning-by-searching is associated with formal activities and is linked to new knowledge, generating both incremental and radical innovations.

External learning can be divided into the following categories:

- learning-by-imitating takes place when innovations introduced by other companies or institutions are introduced in the company;
- learning-by-interacting takes place by interaction and comes as a result of the fact that agents (suppliers, companies, consumers) allow the exchange of information, codes and procedures that result in changes in the status quo of products and processes;
- learning-by-cooperating is developed with other companies, universities, research centers, and other organizations.

Innovative dynamics involve not only the companies' learning mechanisms, but also external agents with or without formal cooperative bonds. Therefore, the innovation concept is related to a wider economic process that involves dynamic relationships between agents involved in the productive process. Innovations may be radical or incremental. Radical innovation, such as development and introduction of a new product or process, may be considered an entirely new way of organizing production. This type of innovation may represent a structural change from the previous technological pattern. It also leads to cost reductions and an increase in the quality of an already existing product. Incremental innovations must recognize improvements in a product, process or organization of a company's production, but with no fundamental changes in the industrial structure (Lemos, 2000). In our study of poultry clusters, incremental innovations are the most observed form.

Innovations have uncertainty since they are the solution to an existing problem and its consequences are not known beforehand (Lemos, 2000). On the other hand, they reveal a cumulative effect since the company's capability to perform changes and advancements is highly influenced by technologies used in the present and by the experience accumulated from the past. For Possas (1991) the technological regime of a company is defined through the relationship with the agro-business market. This relationship plus knowledge and learning processes define possibilities and opportunities for innovation in the form of profit by economic agents.

CONCENTRATION OF POULTRY AGRO-INDUSTRIES IN THE WESTERN REGION OF THE STATE OF PARANA

Since the 1970s, we have seen major changes in the agricultural structure based on techniques brought about by the 'green revolution'. Rather than being focused on older production methods and culture, the economy after 1970 developed into 'technified' agricultural production using modern inputs like soy beans, which has created a new space of distribution of the agricultural production in the state of Parana (Medeiros et al., 2005).

Soy bean production was the major vehicle of regional technical improvement in agricultural production. Cooperatives and agribusinesses were structured around soy bean production. The formation of regional 'soy bean complexes' and corn production has created the base for industrialization of the poultry and pork meat industry in the western region of the state of Parana. Since the end of the 1970s, agro-industrialization of the poultry industry has been an important part of the meat industry production in the region and also for the formation of major oligopolistic companies in Parana and other regions. The introduction of new technology in the production process has allowed vertical integration and production of complementary activities to the slaughter. The poultry industry has grown because of the expansion of soy and corn cultures, thus allowing transforming vegetal proteins into animal proteins (Câmara and Nakazato, 2001; IPARDES, 2003).

Innovations in poultry production in the 1950s and 1960s involved genetics with the development of hybrid poultry specimens. Between 1960 and 1970, there was a sanitary innovation with the development of vaccines and prophylactic techniques. In the 1970s, advancements came from improvements in animal nutrition (Câmara and Nakazato, 2001). In the 1980s, innovation occurred in handling techniques through the installation and use of new equipment. There were significant qualitative changes in the process developments of the industry connected to the acceleration of poultry slaughter technology, resulting in a major upgrade of automation. Also, there were some changes related to the introduction of new types of machinery and equipment in the later steps of the slaughter process, such as new processed products. Changes connected to the innovation strategy of products in later years have come from better control of environment and acclimatization of poultry breeding houses (IPARDES, 2003). Since 2000 we have seen changes occurring in the area of marketing and management involving an increasing differentiation and quality improvement of products (Câmara and Nakazato, 2001).

Consequently, major changes have been connected to emphasis on development of skills and competences acquired through the externalization of production. There were some changes related to growth of agro-industrial and service enterprises which diversified their activities, becoming specialized in their main markets and establishing better relationships with their suppliers, distributors and clients. Also, there were changes connected to the intensification and consolidation of strategic alliances among competitor companies through synergetic activities (Matos and Santos Jr, 1998).

Dalmás et al. (2007) analysed the geographic concentration of poultry agro-industries in the western region of the state of Parana and found strong evidence of the existence of a cluster in regional poultry slaughter and processing sectors. Medeiros (2005) verified that the western and southwest regions of the state of Parana were responsible for 63 percent of the state poultry slaughter in 2004. The poultry sector in both regions was ranked by the study as a concentrated differentiated oligopoly. The companies' competitive situation was characterized by product differentiation and diversification, especially among the export-oriented companies.

The Western Parana region presented the best performance both in generating jobs in the sector (52 percent) and in the growth of poultry production, 85 percent between 1996 and 2006. The poultry slaughter and processing industry is geographically concentrated in the western region of Parana, showing job specialization in the regional economy. These companies also show best performance in relation to the national performance, due mostly to the development of competent export companies.

Brazil had significant participation in worldwide poultry production with 11.93 percent of total global production in 2000, growing to 15.5 percent in 2006. Brazilian poultry exports grew to about 42 percent from 2000 to 2006. Poultry exportation is a complex activity due to the need to keep up the productive efficiency of the whole production phases, updated logistics in the cold storage and transportation and shipment (Gonçalves and Perez, 2006). The southern region of Brazil (where Parana is situated) had 50.2 percent of national poultry production, being responsible for 90 percent of all poultry exports in 2005.

METHODS USED

This study is a research case study of the poultry slaughter and processing cluster located in the western region of the state of Parana. We used a questionnaire that allowed us to collect information that otherwise was unavailable from statistical sources. The questionnaire was based on the Technological Innovation Research (PINTEC) performed by the Brazilian Institute of Geography and Statistics (IBGE).

Our target population consisted of seven poultry slaughter companies with production units located in the western region of the state of Parana which were all members of the Brazilian Association of Poultry Producers and Exporters (ABEF). The identity of the companies will not be revealed. For reporting purposes, they are simply labeled companies A, B, C, D, E, F and G. Of the seven companies, five participated in the research: A, B, C, D and E. The field research took place between December 2007 and April 2008, and the questionnaires were answered by employees connected to the industrial division of each company.

RESULTS AND DISCUSSION

Features of the Companies of the Cluster

The companies taking part in the research project produced several agricultural products, in both domestic and export markets. Three companies were established in the 1960s (A, B, D), one in the 1970s (E) and one in the 1980s (C). The slaughter and processing facilities of companies A, B, C, D and E started in 1997, 1982, 1985, 1995 and 2005 respectively. It can be noted that their production activities came not long after grain production in the region was consolidated. The companies are relatively large and have the following numbers of employees working in the poultry slaughter and processing unit:

A – 3112 employees;
B – 3600 employees;
C – 2300 employees;
D – 2100 employees;
E – 1300 employees.

All the companies buy soy and corn in the western region of the state of Parana on behalf of the primary producers. These data reinforce the idea of a regional cluster because of geographically concentrated labor and raw material supply. Companies belonging to the cluster present a similar trend to the North American slaughter industry, which during the last four decades has increased industrial concentration and production scale (Olinger et al., 2005). The companies studied had no major problems increasing production because of transportation and environmental restrictions, as opposed to

American poultry slaughter and processing industry, as pointed out by Olinger et al. (2005). Regarding the legal constitution of the companies, four were cooperatives (A, B, D, E) and one of the companies was a limited partnership (C). Company C was controlled by a holding company and the other companies (A, B, D, E) were independent companies. The origin of the capital base for these companies was national. Relevant factors to their competitiveness were: price, quality, health and the capability of the productive system to adapt to the market's demands. These factors can be related to the production chain and strong contractual structures between agents which makes them viable in attaining regularity and standardization of raw materials, thus allowing effective exportation planning.

In general, the destination for Brazilian poultry exports consists of a majority of Asian countries, followed by European, Arabic and US markets. The development of exports from the poultry complex in the southern region of Brazil derives from productive efficiency, updated logistics of cold storage, transportation and shipment. Furthermore, the development can be explained by a successful integration between primary poultry producers and the processing and slaughtering industry .

Every company (A, B, C, D, E) had contractual relationships with their suppliers through vertical integration between producer and industry. Company C was the only company working with both integrated and independent producers. The integrating mechanism was fundamental to the development of the poultry sector. Within this structure, the industries offered inputs and technical assistance to the primary poultry producers. On the other hand, the processing plants received commitments from the primary producers to receive poultry with adequate weight before slaughter. The primary producers were responsible for installation and equipment on the farms.

The number of vertically integrated producers among the researched companies was for 2007 distributed as follows:

Company A – 498 integrated primary producers;
Company B – 831 integrated primary producers;
Company C – 390 integrated primary producers;
Company D – 500 integrated primary producers;
Company E – 213 integrated primary producers.

The payment from processing industry to primary producers varied according to the producer's efficiency of handling, food conversion and mortality. According to Tavares and Ribeiro (2007), such types of partnership contracts resulted in higher technological development in poultry production, industrialization, productivity gains, cost reduction, quality and standardization.

In 2007 the amount of poultry slaughtered per day by company was:

A – 224 000 units;
B – 267 000 units;
C – 160 000 units;
D – 140 000 units;
E – 90 000 units.

Innovation Process and Learning

Product innovations

Product innovations are summarized in Table 14.1. Note from this table that companies A and D introduced technologically new products in the market or significantly improved existing products, although those products already existed in the national market. New products to companies A and D included cooked and baked poultry meat (covered with flour) and mechanically recovered meat. Company D developed innovation within its own organization. Company A did this development in cooperation with other companies/institutes. Company D used new raw material/components to launch the innovation, also implying significant changes in the software already incorporated in the company's productive process. This led to higher functionality and more effective use of raw material. For company D, innovation was a product adaptation developed for the export market aimed at complying with the clients' needs. For both companies, innovation generated changes in the productive process such as improvement of the industrial design, functionality and performance, purchase of new machinery or equipment, and introduction of new uses of the raw material significantly differing from how the process earlier was designed. Both companies stated that the new product was not an adaptation of a product developed abroad. Companies B, C and E did not introduce a technologically new or significantly improved product during the period of the research.

Process innovations

Process innovations are summarized in Table 14.2. Note from this table that two companies (A and D) introduced technologically new production processes or significantly improved processes to the company even though such production processes already existed in Brazil. The new production process introduced by company A was based on a process from an industrial facility that produced ready-to-consume poultry. The innovation implemented by company D derived from the imports of machinery to produce recovered meat. In two companies (A and D), the process innovation was based on a process already developed abroad or a radically new production technology. In both companies this innovation demanded purchasing new machinery and equipment and an

Table 14.1 Product innovation in the companies during the 2005–7 period

	A	B	C	D	E
Introduction of technologically new product to the company but already existing in Brazil?	Yes	No	No	Yes	No
Was there introduction of a technologically new product not existing in Brazil?	No	No	No	No	No
Which technologically new product launched by the company in the market?	Products based on cooked, roasted or fried poultry	–	–	Cooked and baked poultry	–
The above described product is:	New to the company but already existing in Brazil	–	–	New to the company but already existing in Brazil	–
Who developed the innovation and where it is located?	The company in cooperation with others – Brazil	–	–	The company – Brazil	–
The major product innovation has incorporated:					
New raw materials?	No	–	–	Yes	–
Improvement in the industrial design?	Yes	–	–	Yes	–
Significant software changing?	No	–	–	Yes	–
Purchase of new machinery?	Yes	–	–	Yes	–
New type of use of the product?	Yes	–	–	Yes	–
Adaptation of foreign product?	No	–	–	No	–
Adaptation of product on client's demand?	No	–	–	Yes	–
Radically new technology?	No	–	–		–

Source: Data from the research

alteration in the order of the production steps. In 2007, only company D had a technologically new or improved project to be implemented in the future. Companies B, C and E did not introduce a technologically new or significantly improved process during the period of the research.

The research analysed innovative activities developed inside the companies. Companies A and D had R&D departments within the company. The importance of the R&D activities performed by company A between 2005 and 2007 was considered low. However, that department was considered important by company A for generating new knowledge. Company D considered the activities performed by the R&D department to be of moderate importance during the period, although it also recognized its importance for generating knowledge.

R&D activities performed by company A were product development, elaboration of new formulations, flavors, textures and spices for the products. R&D activities performed by company D were development of new products, improvement profitability and costs reductions. The R&D department of company D had a kitchen to perform tests (pilot plant) where they defined the parameters of the process for achieving the desired product specifications. Companies B, C and E did not keep internal R&D departments.

R&D activities in companies A and D occurred in a continuous way, and the R&D departments were located in both companies close to the processing plant. Company A had four engineers in the R&D department and company D had two engineers and four food technicians. We found that the R&D departments used two kinds of knowledge: explicit (acquired from books, courses, database or incorporated in equipments and products), and tacit (related to skills in solving problems and to the capability to relate and interact with other people). The creation of organizational knowledge comprised a dynamic and continuous interaction between both types of knowledge.

Companies A, C and D considered the importance of acquiring external knowledge between 2005 and 2007 to be high, companies B and E did not answer these questions. They also considered of high importance the acquisition of machinery, equipment and training performed in the period. The funding sources of the innovative activities performed by companies A and D in the period between 2005 and 2007 consisted of the following: for company A, 30 percent own capital and 70 percent capital from third parties; Company D invested 13 percent of the gross sales of the company in R&D; the remaining amount was collected in the financial market.

Innovations implemented by companies A and D brought different levels of importance to the companies' products, market, processes and routines (Table 14.3). For both companies, innovations were of high importance for the improvement in the quality of the product and to enlarge the range of products offered. For company A, innovations had high importance to the maintenance

Table 14.2 Product process innovation of the companies between 2005 and 2007

	A	B	C	D	E
Was there introduction of a technologically new process to the company but already existing in BR?	Yes	No	No	Yes	No
Was there introduction of a technologically new process not existing in BR?	No	No	No	No	No
Which was the technologically new process introduced by the company?	New industrial facility	–	–	New machinery to produce meat	–
Such process is:	New to the company but already existing in Brazil	–	–	New to the company but already existing in Brazil	–
Who developed such innovation and where is it located?	The company in cooperation with others – Brazil	–	–	Other companies – abroad	–
The major process innovation has incorporated:					
New raw materials?	No	–	–	Yes	–
Purchase of new machinery?	Yes	–	–	Yes	–
Changing in the production setups?	Yes	–	–	Yes	
Organizational changing?	Yes	–	–	No	–
Adaptation of process developed abroad?	No	–	–	No	–
Radically new production technology?	No	–	–	No	–

Source: Data from the research

and growth of the company's participation in the market. For company D, innovations had a medium level of importance. For company A, innovations had high importance for increasing and flexibility in the productive capacity; low importance for production costs, labor and raw material utilization, and they were not relevant to the reduction of water and power consumption. For company D, innovations had low importance for increasing productive capacity; high importance for production flexibility and cost reduction; and medium importance in reducing the work, raw material, power and water costs. For both companies (A and D), innovations had low importance in reducing environmental impacts and aspects related to health and safety, and high importance in regard to complying with regulations, rules and market standards. Companies B and E did not answer these questions.

Results presented in Tables 14.1, 14.2, and 14.3 suggest that innovations implemented by companies often were of the incremental type and they generated improvements in the poultry meat. Such innovations reached the consumer in the form of precooked products, thus responding to market demands. There were no major alterations in the technological standard and therefore no radical innovation. On the other hand, the innovation suggested the cumulative nature of knowledge. The companies' ability to perform changes was strongly influenced by technologies presently used and from experiences accumulated in the past.

Between 2005 and 2007, companies A and D were involved in cooperative arrangements focused on developing innovative activities. Partners of company A involved in developing innovations during that period included clients, suppliers and other companies. Partners of company D included clients, suppliers, other companies connected to the cooperative, consulting companies, universities and research and professional institutes. Companies responded that they did not use governmental support programs (e.g. fiscal incentives and others) for their innovative activities. Between 2005 and 2007, company A used some methods to protect innovations through brand and authorship rights registration. Company D also made a brand registration. Company A had one patent in Brazil and abroad.

The importance attributed to the information sources used between 2005 and 2007 by the companies to develop innovations is presented in Table 14.4. It can be concluded that the innovations must be understood as the result of the learning process that comes from the information flows that circulate between the companies and their environment.

It can be deduced that among different kinds of external learning activities we found learning processes of the learning-by-imitation type (processes based on copying other products and processes already used by other Brazilian and foreign companies). Such learning was performed in an autonomous and non-cooperative way in companies A and D. Also, it verified

Table 14.3 Impact on product and process innovation in companies A and D – grade of importance

Grade of importance	A	D
To the product		
Improvement in the quality of the product	High	High
Amplification in the range of products offered	High	High
To the market		
Maintenance of the participation in the market	High	Medium
Amplification of the participation in the market	High	Medium
Opening of new markets	High	Medium
To the process		
Increase in the productive capacity	High	Low
Increase in the production flexibility	High	High
Reduction of the production costs	Low	High
Reduction of the working costs	Low	Medium
Reduction in the raw material consumption	Low	Medium
Reduction in the power consumption	Non relevant	Medium
Reduction in the water consumption	Non-relevant	Medium
To other impacts		
Reduction in the impact on the environment and controlling of aspects connected to the health and safety	Low	Low
Company compliance to regulations and standard rules of the market	High	High

Source: Data from the research

an external learning-by-interaction process that was due to the interaction between suppliers, the company and its consumers. Such interaction allowed the exchange of information, to set codes and procedures which resulted in the growth of the company's number of products in the market. Learning-by-cooperating occurred more often due to the cooperation with other companies (mainly suppliers) than from universities, research centers and other organizations.

According to Fagerberg (1995), interactions involving tacit knowledge are more necessary in sectors associated with the development of complex and specific technologies. In this study we found mainly incremental innovations,

Table 14.4 Importance of information sources between 2005 and 2007

	A	B	C	D	E
Grade of importance of the sources of information					
Internal sources					
R&D department	High	–	–	High	–
External sources					
Other company of the group	–	–	Low	Medium	Medium
Suppliers	High	Medium	Low	Medium	High
Clients	High	Medium	Low	Medium	High
Competition	–	Low	Low	Non-relevant	Medium
Consulting companies	Medium	–	Non-relevant	Low	Low
Educational and research centers					
Universities and institutes	–	Low	Non-relevant	Low	Low
Professional capacitating centers	–	Medium	Non-relevant	Non-relevant	Medium
Test, assay and certification institutions	–	Low	Non-relevant	Low	Low
Other sources of information					
Acquisition of licenses, patents and know-how	High	Low	Irrelevant	Low	Medium
Conferences, meetings and specialized publications	Low	Low	Non-relevant	Medium	Medium
Fairs and expositions	Medium	Medium	Irrelevant	Medium	High
Informal information nets	–	Non-relevant	Medium	Medium	Medium
Internet sites	–	Medium	Non-relevant	Low	Medium

Source: Data from the research

but tacit knowledge also seemed important for the development of this cluster – formed by an intensive labor industry. Innovations were mainly developed by the companies without much contact and influence from metropolitan regions. According to Drabenstott et al. (1999), in the USA in the last three decades, many old urban industrial meat slaughter and processing facilities were closed. Conversely, new and large facilities were opened in rural communities. It can be verified that the industrial facilities of the studied cluster have some similarity with the American industry trend of having industrial facilities far from large urban centers.

All persons who were interviewed responded that they received knowledge from or socialized with professionals from at least one competing company; meeting and making telephone and email contacts with others at fairs, technical courses, lectures and other events in the area. Therefore, it can be stated that there were informal information flows among the cluster's competing agents.

Also, our results suggest that an important information outflowing process between companies was achieved through workers leaving one company of the cluster and later contracted by another. This situation was reported at every occupation level; however, there was more scarcity in production management and maintenance engineering jobs. Circulation involved more professionals in operational functions with medium technical education. As could be expected, when a company rewarded its professionals, changes between companies were less frequent.

Therefore, every learning form constitutes a base of knowledge to the company and the possibility to enlarge the exchange of experiences stimulated by the proximity of those agents. As such, our results point out the existence of spillovers of located knowledge that provides the main justification for innovative activity to be spatially concentrated or clustered.

FINAL CONSIDERATIONS

Social-institutional structure, intercompany information flowing and learning process are important contributors to the innovation process and company growth. This exists in addition to the benefits derived from developing local potentialities and creating interrelationships between geographically close companies forming a cluster. The results suggested that communication channels present in such arrangements facilitate the diffusion of knowledge, allowing companies located at that site some advantages in innovative performance.

Among the researched companies, results further suggest the existence of external economies that can be looked at as a consequence of the presence of three features:

- existence of a specialized labor contingent;
- presence and attraction of a set of specialized supplier of raw material, components and services;
- dissemination of knowledge, skills and information.

We did not detect external economies coming from intentionally implemented actions by the companies.

Companies also used both explicit type knowledge and tacit knowledge for the development of their business. The companies who performed an innovation during the period used both internal and external learning sources. In the internal innovative process, we identified learning-by-using, learning-by-doing and learning-by-searching. Learning-by-searching was rarely observed despite being recognized as a very important innovation source. The external learning types we identified were learning-by-imitation, learning-by-interaction and learning-by-cooperating. This last process was practically absent due to its incidental nature.

The research revealed that companies have learned new innovative processes both internally and with their competitors by means of informal contacts, transferring a professional to a competitor, and through engineering advances. There is evidence of spillovers from located knowledge, although this was not clearly seen in the learning-by-cooperating process.

Companies within the cluster introduced improvements in product through fully processed poultry meat in the form of precooked products. This kind of innovation revealed a cumulative learning profile, having in mind that their ability to perform those changes received influence from the features of current technologies and from the experience accumulated in the past. However, there were no alterations in the technological standard in use and it implemented no radical innovation, which is not very probable in this kind of industry where innovations are normally of the incremental type.

Thus, companies within the poultry cluster in the western region of the state of Parana construct their competitiveness mainly through interactions with local capacities. They are molded by historic processes and are based on the existing substructure, agricultural structure of small properties, geographically accessible natural resources, institutional structure of the region, as well as in the knowledge, learning and innovation they develop.

REFERENCES

Arrow, K.J. (1969), 'The economic implications of learning by doing', *Review of Economic Studies*, **29**, 155–73.

Besen, G.M.V. and P.R. Delgado (2005), 'Arranjos produtivos locais no parana', *Revista Paranaense de Desenvolvimento*, **109** (jul/dez), 161–4.

Câmara, M.R.G. and R. Nakazato (2001), 'Estratégias competitivas inovadoras em empresas do sistema agroindustrial de frangos no Parana', *Semina: Ciencias Sociais e Humanas*, **22** (1), 23–34.

Dalmás, S.R.S, J.A.R. Staduto and E.M. Willers (2007), 'Da fronteira agrícola a fronteira agroindustrial: Uma análise da concentração das empresas de abate e de processamento da crane de frango no Oeste do Parana', *Revista de Desenvolvimento Econômico*, **9** (16), 48–60.

Drabenstott, M., M. Henry and K. Mitchell (1999), 'Where have all packing plants gone? The new meat geography in rural America', *Economic Review – Federal Reserve Bank of Kansas City*, Third Quarter, 65–81.

Fagerberg, J. (1995), 'User-producer interaction learning and comparative advantage', *Cambridge Journal of Economics*, **19**, 243–56.

Gonçalves, J.S. and L.H. Perez (2006), 'Exportações Brazileiras da cadeia de produção de aves no período 2000–2005: origem, destino e agregação de valor', *Informações Econômicas*, **36** (7), 32–47.

IPARDES (Instituto Paranaense de Desenvolvimento Econômico e Social) (2003), *Leituras Regionais: Microrregião Geográfica Oeste Paranaense*, Curitiba: IPARDES/BRDE.

Jacobs, J. (1969), *The Economy of Cities*, New York: Random House.

Lemos, C. (2000), 'Inovação na era do conhecimento', *Parcerias Estratégicas*, **8**, 157–79.

Malerba, F. (1992), 'Learning by firms and incremental technical change', *The Economic Journal*, **102** (413), 845–59.

Marshall, A. (1890, 8th edn 1920), *Principles of Economics*, London: Macmillan.

Matos, V.A. and W.L. Santos Jr (1998), 'A avicultura Brazileira e suas articulações: uma análise do desenvolvimento dos negócios a partir das alianças estratégicas', Anais, ANPAD – Associação Nacional de Pós-Graduação em Administração.

Medeiros, N.H. et al. (2005), A avicultura nas mesorregiões Oeste e Sudoeste paranaense: uma análise de concentração de mercado, IV Encontro de Economia Paranaense, Anais, Toledo (PR): Edunioeste.

Ollinger, M., J. MacDonald and M. Madison (2005), 'Technological change and economies of scale in US poultry processing', *American Journal of Agriculture Economics*, **87** (1), 116–29.

Possas, M.L. (1991), 'Concorrência, inovação e complexos industriais: algumas questões conceituais', *Cadernos de Ciência & Tecnologia*, **8** (1/3), 78–97.

Romer, P.M. (1986), 'Increasing returns and long-run growth', *Journal of Political Economy*, **94** (51), 1002–37.

Rosenberg, N. (1982), *Inside the Black Box: Technology and Economics*, Cambridge: Cambridge University Press.

Silva, M.V.B. and R.M. Silveira Neto (2007), 'Crescimento do emprego industrial no Brazil e geografia econômica: evidências para o período pós-real', *Revista Economia*, **8** (2), 269–88.

Suzigan, W. et al. (2003), Sistemas locais de produção: mapeamento, tipologia e sugestões de políticas, XXXI Encontro Nacional de Economia, Porto Seguro, Anais, Porto Seguro: ANPEC.

Tavares, L.P. and K.C. Ribeiro (2007), 'Desenvolvimento da avicultura de corte Brazileira e perspectivas frente à influenza aviária', *Organizações Rurais & Agroindustriais*, **9** (1), 79–88.

Index

Note: Page numbers in italics represent tables and figures.